10 0614198 4

KU-000-611

UNI

FROM THE LIBRARY

Advances in
CHEMICAL ENGINEERING

MICROSYSTEMS AND DEVICES FOR

(BIO)CHEMICAL PROCESSES

VOLUME **38**

ADVANCES IN
CHEMICAL ENGINEERING

Editor-in-Chief

GUY B. MARIN
Department of Chemical Engineering
Ghent University
Ghent, Belgium

Editorial Board

DAVID H. WEST
Research and Development
The Dow Chemical Company
Freeport, Texas, U.S.A.

JINGHAI LI
Institute of Process Engineering
Chinese Academy of Sciences
Beijing, P.R. China

SHANKAR NARASIMHAN
Department of Chemical Engineering
Indian Institute of Technology
Chennai, India

Advances in
CHEMICAL ENGINEERING

MICROSYSTEMS AND DEVICES FOR
(BIO)CHEMICAL PROCESSES

VOLUME **38**

Edited by

J C SCHOUTEN

Eindhoven University of Technology
Eindhoven, The Netherlands

UNIVERSITY OF NOTTINGHAM
GEORGE GREEN LIBRARY OF
SCIENCE AND ENGINEERING
WITHDRAWN
FROM THE LIBRARY

Amsterdam • Boston • Heidelberg • London • New York • Oxford
Paris • San Diego • San Francisco • Singapore • Sydney • Tokyo
Academic Press is an imprint of Elsevier

ELSEVIER

Academic Press is an imprint of Elsevier
Linacre House, Jordan Hill, Oxford OX2 8DP, UK
32 Jamestown Road, London NW1 7BY, UK
Radarweg 29, PO Box 211, 1000 AE Amsterdam, The Netherlands
30 Corporate Drive, Suite 400, Burlington, MA 01803, USA
525 B Street, Suite 1900, San Diego, CA 92101-4495, USA

First edition 2010

Copyright © 2010, Elsevier Inc. All rights reserved

No part of this publication may be reproduced, stored in a retrieval system
or transmitted in any form or by any means electronic, mechanical, photocopying,
recording or otherwise without the prior written permission of the publisher

Permissions may be sought directly from Elsevier's Science & Technology Rights
Department in Oxford, UK: phone (+44) (0) 1865 843830; fax (+44) (0) 1865 853333;
email: permissions@elsevier.com. Alternatively you can submit your request online by
visiting the Elsevier web site at http://www.elsevier.com/locate/permissions, and selecting:
Obtaining permission to use Elsevier material

Notice
No responsibility is assumed by the publisher for any injury and/or damage to persons
or property as a matter of products liability, negligence or otherwise, or from any use
or operation of any methods, products, instructions or ideas contained in the material
herein

ISBN: 978-0-12-374458-6
ISSN: 0065-2377

For information on all Academic Press publications
visit our website at elsevierdirect.com

Printed and bound in USA
10 11 12 13   10 9 8 7 6 5 4 3 2 1

Working together to grow
libraries in developing countries

www.elsevier.com | www.bookaid.org | www.sabre.org

ELSEVIER   BOOK AID International   Sabre Foundation

CONTENTS

Anıl Ağıral, *Mesoscale Chemical Systems, MESA + Institute for Nanotechnology, University of Twente, 7500 AE Enschede, The Netherlands*

Arata Aota, *Institute of Microchemical Technology, and Micro Chemistry Group, Kanagawa Academy of Science and Technology, 3-2-1 Sakado, Takatsu, Kawasaki, Kanagawa 213-0012, Japan*

A.J. deMello, *Department of Chemistry, Imperial College London, South Kensington, London SW7 2AY, UK*

J. C. deMello, *Department of Chemistry, Imperial College London, South Kensington, London SW7 2AY, UK*

Han J.G.E. Gardeniers, *Mesoscale Chemical Systems, MESA + Institute for Nanotechnology, University of Twente, 7500 AE Enschede, The Netherlands*

Takehiko Kitamori, *Department of Applied Chemistry, School of Engineering, The University of Tokyo, 7-3-1 Hongo, Bunkyo, Tokyo 113-8656, Japan, and Micro Chemistry Group, Kanagawa Academy of Science and Technology, 3-2-1 Sakado, Takatsu, Kawasaki, Kanagawa 213-0012, Japan*

S. Krishnadasan, *Department of Chemistry, Imperial College London, South Kensington, London SW7 2AY, UK*

Paul Watts, *Department of Chemistry, The University of Hull, Cottingham Road, Hull HU6 7RX, UK*

Charlotte Wiles, *Department of Chemistry, The University of Hull, Cottingham Road, Hull HU6 7RX, UK; Chemtrix BV, Burgemeester Lemmensstraat 358, 6163JT Geleen, The Netherlands*

A. Yashina, *Department of Chemistry, Imperial College London, South Kensington, London SW7 2AY, UK*

This volume of *Advances in Chemical Engineering* has "microsystems and devices for (bio)chemical processes" as its central theme. Four chapters are presented that cover different aspects of this theme, ranging from continuous flow processing in microsystems to nanoparticle synthesis in microfluidic reactors. This field of microprocess technology has experienced a very large growth during the last two decades as is reflected by the even still today continuously increasing number of scientific journal publications, reviews, patents, text books, and monographs. Significant developments in the field of miniaturization of lab-scale systems and even complete plants for chemicals and materials production have been achieved. Microreactors and microfluidic devices are now being used at the industrial level in widely diverse areas, such as fuel processing, fine chemicals synthesis, functional material synthesis, high-throughput catalyst screening, polymerization, and sensors and process analytics. The application field is still expanding as chemists and engineers more and more acknowledge the benefits of taking advantage of the microscale in the efficient and controlled production of chemicals and materials at so far often unprecedented operating conditions.

The advantages of continuous operation at the microscale in chemicals and materials production are well described in the literature. Many examples are known in which the high rates of mass and heat transfer under well-defined and well-controlled fluid flow regimes contribute significantly to improved conversions and yields. As a result, also new process windows are explored and new chemistries are discovered. This volume of *Advances in Chemical Engineering* addresses a few of these developments. In the first chapter, Arata Aota and Takehiko Kitamori discuss the need for general concepts for the integration of microunit operations in microchemical systems that are used for continuous flow processing. They focus in particular on analysis, synthesis, and construction of biochemical systems on microchips, which show superior performance in the sense that these systems provide rapid, simple, and highly efficient processing opportunities.

In the second chapter, Anil Ağiral and Han J.G.E. Gardeniers take us to a fascinating world wherein "chemistry and electricity meet in narrow alleys." They claim that microreactor systems with integrated electrodes provide excellent platforms to investigate and exploit electrical principles as a means to control, activate, or modify chemical reactions, or even preparative separations. Their example of microplasmas shows that the chemistry can take place at moderate temperatures where the reacting species still have a high reactivity. Several electrical concepts are presented and novel principles to control adsorption and desorption, as well as the activity and orientation of adsorbed molecules are described. The relevance of these principles for the development of new reactor concepts and new chemistry is discussed.

The third chapter by Charlotte Wiles and Paul Watts addresses high-throughput organic synthesis in microreactors. They explain that one of the main drivers for the pharmaceutical industry to move to continuous production is the need for techniques which have the potential to reduce the lead time taken to generate prospective lead compounds and translate protocols into production. The rapid translation of reaction methodology from microreactors employed within R&D to production, achieved by scale-out and numbering-up, also has the potential to reduce the time needed to take a compound to market. The authors discuss many examples of liquid phase, catalytic, and photochemical reactions and they conclude the chapter with a selection of current examples into the synthesis of industrially relevant molecules using microreactors.

The last chapter by A.J. deMello and his coworkers gives an overview of microfluidic reactors for nanomaterial synthesis. The authors explain that the difficulty of preparing nanomaterials in a controlled, reproducible manner is a key obstacle to the proper exploitation of many nanoscale phenomena. In the chapter, they describe recent advances in the development of microfluidic reactors for controlled nanoparticle synthesis. In particular, recent work of their group aimed at developing an automated chemical reactor capable of producing on demand and at the point of need, high-quality nanomaterials, with optimized physicochemical properties, is highlighted. This automated reactor would find applications in areas such as photonics, optoelectronics, bioanalysis, targeted drug delivery, and toxicology where it is essential to characterize the physiological effects of nanoparticles not only in terms of chemical composition but also size, shape, and surface functionalization.

The developments described in these four chapters are all part of what we call today "process intensification" which aims at process equipment innovation, enabling new manufacturing and processing methods. This has led already to interesting foresight scenarios in which it is envisaged that future leading process technology will be based on a widespread implementation and use of intensified, high-precision process equipment and devices, including corresponding adaptation of plant management, supply chain organization, and business models. New reactor engineering options are foreseen with considerably improved energy and process efficiencies and economics. Challenging concepts are envisaged such as multipurpose, self-adapting, and modular process devices using advanced sensor and process analyzer technologies and programmable chemical reactors, whose local operating conditions adapt automatically to changes in feed composition and product specifications. Important drivers for introduction of these new equipment technologies are market flexibility with shorter delivery times, decentralized and continuous manufacturing, and just-in-time and on-demand production closer to the end-user. It is evident that these developments will strongly impact on the nature and scale of process equipment, pilot plants, and production facilities. Here microreactors and microfluidic devices will play an important role as well.

I hope you find the chapters in this volume of interest to your work. Of course, this volume does not cover all topics in the field of microprocess technology and microfluidics. For example, recent developments in the field of catalytic coating development or the use of alternative energies such as microwaves or spinning action have not been covered. These new developments will definitely be the subject of upcoming reviews and exemplify the ongoing research in a very challenging area.

Jaap Schouten
Eindhoven University of Technology,
Eindhoven, the Netherlands

CHAPTER 1

Microunit Operations and Continuous Flow Chemical Processing

Arata Aota[1] and **Takehiko Kitamori**[2,*]

Contents

Abstract

Integrated microchemical systems on microchips are expected to become important tools for analysis and synthesis within the biological sciences and technologies. For these purposes, general integration concepts were developed, including microunit

1   Institute of Microchemical Technology, and Micro Chemistry Group, Kanagawa Academy of Science and Technology, 3-2-1 Sakado, Takatsu, Kawasaki, Kanagawa 213-0012, Japan

2   Department of Applied Chemistry, School of Engineering, The University of Tokyo, 7-3-1 Hongo, Bunkyo, Tokyo 113-8656, Japan, and Micro Chemistry Group, Kanagawa Academy of Science and Technology, 3-2-1 Sakado, Takatsu, Kawasaki, Kanagawa 213-0012, Japan

* Corresponding author.
E-mail address: mailto:kitamori@icl.t.u-tokyo.ac.jp

Advances in Chemical Engineering, Volume 38
ISSN: 0065-2377, DOI 10.1016/S0065-2377(10)38001-X

© 2010 Elsevier Inc.
All rights reserved.

operations and continuous-flow chemical processing. The general methodology has enabled analysis, synthesis, and construction of biochemical systems on microchips, and these microsystems have demonstrated superior performance (i.e., rapid, simple, and highly efficient processing). Microchemical technology has now entered the phase of practical application. In this chapter, we discuss general methods, and applications of microchemical systems on microchips.

1. INTRODUCTION

Integrated microchemical systems are essential tools for high-speed, functional, and compact instrumentations used for analysis and synthesis, biological sciences, and technologies. In the 1990s, most microchip-based systems were used for gene or protein analysis and employed electrophoretic separation with laser-induced fluorescence detection (Auroux et al., 2002; Reyes et al., 2002). However, other analytical and synthesis methods were required for more general analytical, combinatorial, physical, and biochemical applications that involve complicated chemical processes, organic solvents, neutral species, and nonfluorescent molecule detection. For these applications, general microintegration methods on microchips became quite important.

General methods for microintegration of chemical systems that are similar to systems used in electronics have been developed (Figure 1). Instead of the resistor, capacitor, and diode, mixing, extraction, phase separation, and other unit operations of chemical processes are integrated as components. These unit operations are known as microunit operations (MUOs), and like the parts of an electric circuit. They can combine with one another in parallel and in serial by continuous-flow chemical processing (CFCP). The function of the microchemical chip is analogous to that of a chemical central processing unit. In order to realize these basic concepts, fluidic control methods are quite important, and parallel multiphase microflows were realized by partial surface modification and channel structures which enabled various MUOs and flexible integration of the MUOs connected by CFCP. This technique has demonstrated superior performance in shorter processing times (from days or hours to minutes or seconds), smaller sample and reagent volumes (down to a single drop of blood), easier operation (from professional to personal), and smaller system sizes (from 10 m chemical plants to desktop plants to mobile systems) compared to conventional analysis, diagnosis, and chemical synthesis systems. Practical prototype systems have also been realized in environmental analysis, clinical diagnosis, cell analysis, gas analysis, medicine synthesis, microparticle synthesis, and so on. For example, a

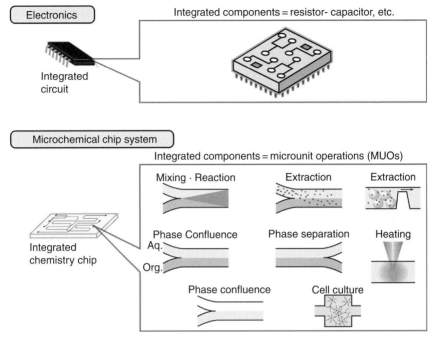

Figure 1 A microchemical chip system compared with an electronic system.

portable microELISA (enzyme-linked immunosorbent assay) system was recently constructed where the analysis time was several minutes, rather than several hours as in the conventional technique. The sample volume (5 μl) was also considerably smaller than the volume (several milliliters) required for conventional diagnosis. In addition, the operation itself was much easier, and its correlation with conventional methods was confirmed with real blood samples. These features are promising for point-of-care (POC) clinical diagnosis.

Microtechnology is moving in two directions. The first is the practical application. Although some practical prototype systems have been realized, attaining long-term stability in parallel liquid/liquid and liquid/gas microflows can be sometimes problematic, and the interfaces are distorted. For robust fluidic control, methods for stabilizing or recovering parallel multiphase microflows are essential. For microfluidic devices, the ability to process small (nanoliter to microliter volume) samples and to interface with microchips is important. In addition, smart fluidic devices (valves or pumps) for flexible and reliable liquid handling on microchips are required for more complex and high-throughput chemical processing on microchip.

The second direction involves new science in nanoscale (10–1,000 nm) space, called as extended nanospace. The extended nanospace is an important region in bridging the gap between single molecules and the condensed phase. Recently, general methodologies in microspace were applied to extended nanospace demonstrating that the basic methodologies in microspace are applicable to extended nanospace. Water displays remarkably different properties in the extended nanospace channel. A model fundamentally related to the surface chemistry of the nanochannel was proposed. This model was applied to protein analysis of a countable number of molecules, and the same type of integrated approach (extended nanospace channels embedded in microfluidic systems) was applied to control and analysis of cell immobilization and culture.

Here, we introduce these basic methodolodgies and applications. Firstly, the general concepts (MUO and CFCP) for microintegration in Chapter 2, and fluid control methods to support the MUO and CFCP in Chapter 3 are discussed.

2. DESIGN AND CONSTRUCTION METHODOLOGY FOR INTEGRATION OF MICROCHEMICAL SYSTEMS

Integrated microchemical systems are expected to be applied in various fields. In order to realize general analytical, combinatorial, physical, and biological applications, general microintegration methods on microchips are quite important. Conventional macroscale chemical plants or analytical systems are constructed combining unit operations such as mixer, reactors, and separators. Similar methodology can be applied to the microchemical systems. By combining MUOs with different functions in series and in parallel, various chemical processes can be integrated into microchips through use of a multiphase microflow network such as that shown in Figure 2. This methodology is termed CFCP (Tokeshi et al., 2002). Miniaturizing conventional unit operations is often not effective and sometimes does not work at all, because many physical properties (e.g., heat and mass transfer efficiency, specific interfacial area, and vanishingly small gravity force) are significantly different in microspace. Therefore, novel MUOs taking these issues into account are needed. Kitamori and colleagues have developed MUOs for various purposes, such as mixing and reaction (Sato et al., 1999; Sorouraddin et al., 2000, 2001), phase confluence and separation (Aota et al., 2007a, 2007b; Hibara et al., 2002, 2003; Hisamoto et al., 2001a, 2001b, 2003; Hotokezaka et al., 2005; Kikutani et al., 2004; Minagawa et al., 2001; Miyaguchi et al., 2006; Sato et al., 2000b; Smirnova et al., 2006, 2007; Surmeian et al., 2001, 2002; Tokeshi et al., 2000a, 2000b), solvent extraction (Aota et al., 2007a, 2007b; Hibara et al., 2001, 2003; Hisamoto et al., 2001a, 2001b; Hotokezaka

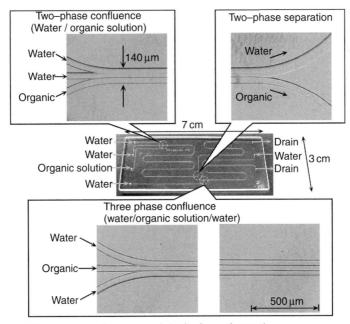

Figure 2   Multiphase microflow network (Tokeshi et al., 2002).

et al., 2005; Kikutani et al., 2004; Minagawa et al., 2001; Miyaguchi et al., 2006; Sato et al., 2000b; Smirnova et al., 2006, 2007; Surmeian et al., 2001, 2002; Tokeshi et al., 2000a, 2000b, 2002), gas–liquid extraction (Aota et al., 2009b; Hachiya et al., 2004), solid-phase extraction and reaction on surfaces (Kakuta et al., 2006; Ohashi et al., 2006; Sato et al., 2000a, 2001, 2003, 2004), heating (Goto et al., 2005; Slyadnev et al., 2001; Tanaka et al., 2000), cell culture (Goto et al., 2008; Tamaki et al., 2002; Tanaka et al., 2004, 2006), and ultrasensitive detection (Hiki et al., 2006; Mawatari et al., 2006; Proskurnin et al., 2003; Tamaki et al., 2002, 2003, 2005; Tokeshi et al., 2001, 2005; Yamauchi et al., 2006) (Figure 3).

2.1 Multiphase microflow network

For many chemical processing applications, microchemical systems should include solvent extraction and interfacial reaction components utilizing both aqueous and organic (or gas and liquid) solutions. Both solutions must be controlled to realize general chemistry in a microchip. In 1990s, electroosmotic flow was used in microchip electrophoresis (Auroux et al., 2002; Reyes et al., 2002); however, the electroosmotic flow is restricted to the flow control of only one type solution (aqueous buffer). Therefore, electroosmotic flow is not suitable for a flow-control method to

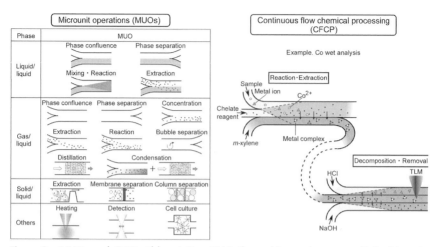

Figure 3   MUOs and CFCP. Abbreviation: TLM, thermal lens microscope (Tokeshi et al., 2002).

integrate various chemical processes that require other types of solvents. Simple pressure-driven flow has been used by many researchers as a nonelectrical pumping method (Brody and Yager, 1997; Kopp et al., 1998; Weigl and Yager, 1999).

Recently, microsegmented flows of immiscible solutions were used for microchemical processes (Burns and Ramshaw, 2001; Kralj et al., 2005; Sahoo et al., 2007; Song et al., 2003). Although the microsegmented flows are advantageous in mixing and rapid molecular transport, phase separation and more than three-phase contact are difficult. Furthermore, controlling the microsegmented flows in the wide range of the flow rates is difficult because the size of the segments changes along with the flow rate ratio. Kitamori et al. have developed parallel multiphase microflows, which flow side-by-side along the microchannels (Hibara et al., 2001). Parallel multiphase microflows in laminar flow regime allow better design and control of microchemical processes because the parallel multiphase microflows are stable in a wide range of the flow rates and can create more than three-phase contact.

2.2 Example of microchemical processes

2.2.1 Molecular transport in microspace

Microfluidic chemical processes are based on transverse molecular diffusion to the microchannel. In microspace, because the diffusion distance is short, rapid chemical processes are expected there. In order to clarify the time required for chemical processes in microspace, diffusion time t in one dimension is considered for a simple discussion. t in one dimension is characterized as follows:

$$t = \frac{l_d^2}{K} \tag{1}$$

where l_d and K are the diffusion length and the diffusion coefficient. Figure 4 shows the diffusion time dependence on the diffusion length. Solid and dashed lines correspond to the diffusion time when the respective diffusion coefficients are 1×10^{-9} and $1 \times 10^{-10}\,\mathrm{m^2\,s^{-1}}$. In 100-μm-wide microchannels, diffusion time is just 10 s for a sample having a diffusion coefficient of $1 \times 10^{-9}\,\mathrm{m^2\,s^{-1}}$. Therefore, samples of high diffusion coefficients, for example, metal chelates, are rapidly transported in microspace. However, diffusion time is 100 s for a sample having a diffusion coefficient of samples of $1 \times 10^{-10}\,\mathrm{m^2\,s^{-1}}$. Samples of low diffusion coefficient, for example, DNA and proteins, cannot be quickly transported in microspace.

2.2.2 Co wet analysis

Micro cobalt wet analysis is an example of CFCP that has been demonstrated (Figure 5) (Tokeshi et al., 2002). Conventional procedures for such analysis consist of a chelating reaction, a solvent extraction of the complex, and a decomposition and removal of coexisting metal complex. These procedures are analogous to the conventional unit operations; mixing and reaction, phase confluence, solvent extraction, phase separation, phase confluence, and solvent extraction. These unit operations, in the same way, are analogous to the MUOs. CFCP is designed by combining the MUOs in series and in parallel, as shown in Figure 3.

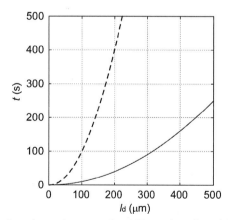

Figure 4 Diffusion time dependence on the diffusion length. Solid and dashed lines correspond to the diffusion time when the respective diffusion coefficients are 1×10^{-9} and $1 \times 10^{-10}\,\mathrm{m^2\,s^{-1}}$.

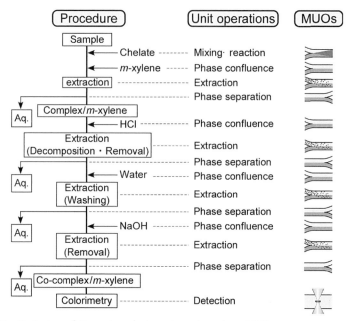

Figure 5   Designing of Co wet analysis systems based on CFCP.

The microchip consists of two different areas, one for reaction and extraction and the other for purification (i.e., decomposition, extraction, and removing impure chelates). In the reaction area, the sample solution containing Co(II), 2-nitroso-1-naphthol (NN), and *m*-xylene were introduced at a constant flow rate through three inlets using microsyringe pumps. These three flows converge at the intersection point, and parallel two-phase microflows, consisting of an organic–aqueous interface, forms in the microchannel. The chelating reaction of Co(II) and NN, and extraction of the Co–NN chelates proceed as the reacting mixture flows along the microchannel. Because NN reacts with coexisting metal ions, such as Co (II), Ni(II), and Fe(II), these coexisting metal chelates are also extracted into the *m*-xylene microflow. Therefore, purification is needed after extraction.

The coexisting metal chelates are decomposed when they make contact at the HCl–*m*-xylene interface, and the metal ions are dissolved in the HCl solution phase. The decomposed fragment of NN is dissolved in NaOH solution, and the Co–NN chelate is stable in concentrated HCl and NaOH solutions, where it remains. Finally, the target chelates in *m*-xylene are detected by a thermal lens microscope downstream, where the Co(II) in aqueous solutions was successfully determined. The limit of detection (2σ) was 0.13 zmol or 78 chelate molecules. The analysis time in this system is only 50 s versus 6 h for conventional devices. Micro chemical processing through use of CFCP has been demonstrated on numerous occasions (Table 1).

Table 1 Example of microchemical processing

Method	Analyte	Limit of detection	Analysis time	Reference
Solvent extraction	Iron complex	7.7 zmol	60 s	Tokeshi et al. (2000)
	Cobalt complex	0.13 zmol, 0.72 zmol, 0.072 zmol	60 s, 50 s, 0 min	Tokeshi et al. (2000, 2002), Minagawa et al. (2001)
	Nickel complex	NA	5 min	Sato et al. (2000)
	K$^+$ and Na$^+$	4.5 zmol	1 s	Hisamoto et al. (2001a, 2001b)
	Dye molecules	NA, NA	6 s, 4 s	Surmeian et al. (2001, 2002)
	Amphetamines	0.5 µg ml^{-1}	15 min	Miyaguchi et al. (2006)
	Carbaryl	0.5 zmol, 0.079 zmol	4.5 min, 5 min	Smirnova et al. (2006, 2007)
	Uran (VI)	0.86 amol	1 s	Hotokezaka et al. (2005)
Gas extraction	Formaldehyde	8.9 ppb	30 min	Hachiya et al. (2004)
	Ammonia	1.4 ppb	16 min	Aota et al. (2009b)
Immunoassay	S-IgA	<1 µg ml^{-1}	<1 h	Sato et al. (2000)
	Carcinoembryonic antigen	30 pg ml^{-1}	35 min	Sato et al. (2001)
	Interferon-γ	10 pg ml^{-1}	50 min (four samples)	Sato et al. (2003)

Table 1 *(Continued)*

Method	Analyte	Limit of detection	Analysis time	Reference
	B-type natriuretic peptide	$0.1\,pg\,ml^{-1}$	35 min	Sato et al. (2004)
	IgE	$2\,ng\,ml^{-1}$	12 min	Kakuta et al. (2006)
	Prion	$50\,pg\,ml^{-1}$	15 min	–
	C-reactive protein	$20\,ng\,ml^{-1}$	8 min	–
	Amphetamines	$0.1\,ng\,ml^{-1}$	10 min	–
Flow Injection	Fe^{2+}	6 zmol	150 s	Sato et al. (1999)
	Ascorbic acid	1 zmol	30 s	Sorouraddin et al. (2000)
	Catecholamines	2 zmol	15 s	Sorouraddin et al. (2001)
Enzymatic assay	H_2O_2	NA, 50 zmol, NA	250 s, NA, NA	Hisamoto et al. (2003), Tanaka et al. (2000, 2004)

The CFCP approach can be applied to develop more complicated processing system. More rapid analysis, bioassays, and immunoassays, as well as more efficient reaction and extraction, can be achieved through CFCP systems as compared to conventional devices. However, development of complicated microchemical processing systems is time consuming because of the need to determine appropriate conditions for each reaction and separation process at the miniaturized scale, and also design and optimize the microchannel structure. In semiconductor circuit device design, computer-aided design (CAD) is typically used to perform circuit analysis, integrated device analysis, layout design, and logical simulation. If CAD is developed specifically for microchemical processing, which includes microfluid simulation, reaction time analysis, extraction time analysis, and microchannel structure design, the time of development for these systems would be drastically shortened.

2.2.3 Microimmunoassay

Immunoassay is one of the most important analytical methods for clinical diagnosis. Heterogeneous immunoassay is a particularly popular approach, wherein captured antibodies or antigens fixed on a solid surface react with serum target antigens or antibodies. Labeled secondary antibodies then capture the targets, and the target amounts are quantified by detecting the labels. Heterogeneous immunoassay features easy and accurate bound-free separation, which is important for improving limit of detection (LOD). Shortening analysis time is necessary when dealing with smaller samples and reagent volumes, given increasing demand in POC diagnosis. In conventional methods, a microtiter plate well (typically 0.65 mm in diameter) is used, and the specific interfacial area for 50 μl volumes is 13 cm^{-1}. However, when a microchannel (200 μm deep and 200 μm wide) is utilized for immobilizing the antibodies (antigens), the interfacial area, which affects the reaction time and efficiency, increases to 200 cm^{-1}.

Kitamori et al. have reported the first successful integration of ELISA into a microchip (Kakuta et al., 2006; Sato et al., 2000a, 2001, 2003, 2004). They developed a new MUO for the heterogeneous reaction through the use of bead-packed microchannels (Figure 6). They immobilized antibodies (antigens) on the bead surface via physisorption or chemisorption. After immobilization, the authors applied a bovine serum albumin or MPC (2-methylacryloyloxyethyl phosphorylcholone) polymer coating to minimize nonspecific adsorption. A microbead suspension with captured antibodies was introduced and stopped at a microchannel dam structure (with a 10 μm gap), and a 5–10-mm-long area of packed beads area was formed. The target sample solution was introduced and captured by the antibodies (antigens). Enzyme-conjugated secondary (horseradish peroxidase (HRP) or alkaline phosphatase (ALP)) antibodies were then introduced to capture the antigens. Finally, substrates were applied, and the dye molecules

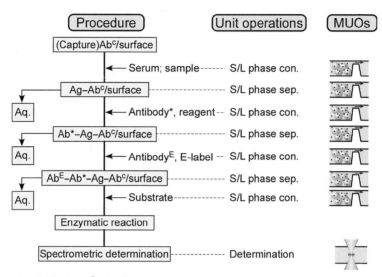

Figure 6 Designing of microimmunoassay.

produced by the enzymatic reaction were detected by thermal lens microscope (TLM), downstream from the microbeads.

Instrumentation for easy operation and high precision is also an important for clinical diagnostic systems. Usually, fluidic handling in microchannels is complex and difficult due to the lack of low volume (1 μl) valves or fluidic interfaces from macroscale to microscale without generating dead volume of microliter scale. Furthermore, it is advisable that microchips be replaced when new measurements are taken, or laborious procedures to remove antigens and wash the microchannels are required. These factors limit the expansion of microchip technologies for clinical diagnosis. The developed method features the utilization of microbeads for immunoreactions; and by simply introducing new microbeads, new measurements can easily be taken. Recently, Kitamori et al. developed a portable micro-ELISA system for POC clinical diagnosis by designing a simple microfluidic system with integrated fluidics, optics, and electronics (Ohashi et al., 2006). An example of our system and the components for four parallel measurements are illustrated in Figure 7. The samples and reagents are stocked in tubes on a moving tray. A PEEK™ tube, under the control of four syringe pumps, aspirates and injects the samples and reagents (microliter volume) into the microchip. The diversion of flow, to either waste or microchip, is controlled by their novel, low volume (~1 μl) and fast response time (<1 s), microvalve. The tube is connected to the microchip through a hydrophobic connector with pressure resistance of >1 MPa. Under a maximal flow rate of 10 μl min^{-1} during washing process, the pressure drop at the bead-packed area is <100 kPa. Immunoassay is then conducted on the microchip and detected

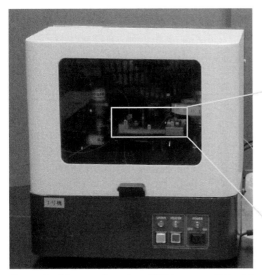
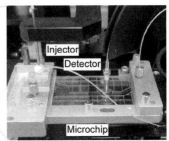

Figure 7   Picture of a micro-ELISA system (Ohashi et al., 2006).

by an on-chip TLM (Tokeshi et al., 2005). In this case, stopped-flow conditions were utilized for the enzymatic reaction, and signals were obtained as a clear peak for each channel. All the processes are controlled by software on a PC.

2.2.4 Stimulant analysis in urine

Urine analysis for illegal drugs is increasingly performed in forensic laboratories (especially in Japan). Gas chromatography–mass spectrometry (GC–MS) is extensively used because of its versatility and reliability. By way of sample preparation for GC analysis, conventional liquid–liquid extraction has a widespread use, but it is not only laborious but also environmentally unfriendly due to the consumption of considerable amounts of organic solvents. Therefore, microintegration of the sample preparation procedure is required.

Conventional procedures and corresponding MUOs are illustrated in Figure 8 (Miyaguchi et al., 2006). Urine samples are mixed with organic solvent (1-Chlorobutane) and stimulants in urine are extracted. After phase separation, the organic solution is mixed with derivatization reagent (trifluoroacetic anhydride). Unreacted derivatization reagents were washed by extracting with water. The washed organic solution was collected in a vial, and the concentration was determined by GC–MS. These conventional processes are easily dissolved to unit operations and converted to corresponding MUOs. The MUOs can be integrated to microchips by CFCP as shown in Figure 9. The conventional procedures can be simplified by the multiphase microflow network. The sample preparation time was decreased from several

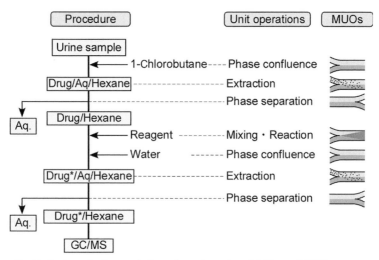

Figure 8　Designing of urine analysis systems based on MUOs and CFCP.

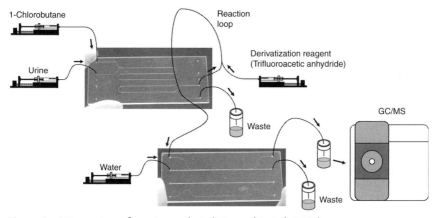

Figure 9　Microsystems for urine analysis (Miyaguchi et al., 2006).

hours to 5 min, and total analysis time including GC–MS is below 20 min. In addition, just 100 μl of sample and reagent are required. These features will open on-site and precise urine analysis in near future.

3. SURFACE CHEMISTRY FOR MULTIPHASE MICROFLOWS

3.1 Fundamental physical properties of multiphase microflows

Control of multiphase microflows is an important basic technology for integration of MUOs. Therefore, control methods for the stable phase separation of the multiphase microflows in a wide range of the flow rates are required. However, the methods in the conventional size devices cannot be applied because the physical properties in the microspace are different from those in macrospace. In conventional devices, the aqueous and organic phases are separated by gravity. In the microspace, however, the fluid is greatly influenced by liquid–solid, liquid–gas, and liquid–liquid interfaces because of the large specific interfacial area. To clarify the main physical forces in the microchannels which include viscous force and interfacial parameter and can be analyzed using the dimensionless Reynolds (Re) and Bond (Bo) numbers, defined as an inertia-to-viscous force ratio and a gravity-to-tension ratio, respectively. Figure 10a shows a cross section of a model microchannel with dimensions included in these parameters.

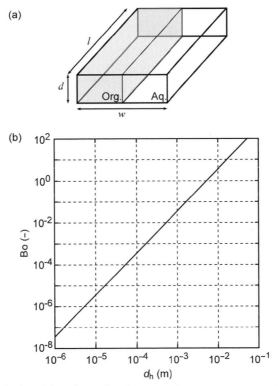

Figure 10 (a) The liquid–liquid interface between the aqueous an organic phases (Aq. and Org.) in a model microchannel. The width, depth, and length along the microchannel are represented as w, d, and l, respectively. (b) Bond number (Bo) dependence on the hydrodynamic diameter for water–toluene microflows.

Bo is defined as

$$Bo = \frac{(\Delta\rho)gd_h^2}{\gamma} \tag{2}$$

where $\Delta\rho$, γ, and d_h are the density difference, the interfacial tension between the two phases, and equivalent diameter, respectively, and where g is the gravitational acceleration ($9.8\,\mathrm{m\,s^{-2}}$). The variable d_h corresponds to the mean hydraulic diameter defined as

$$d_h = \frac{4S}{l_p} \tag{3}$$

where S is the cross section and l_p is the perimeter of a section of the microchannel. Figure 10b shows the Bo of the water–toluene two phase flows as a function of the d_h ($\Delta\rho = 0.132 \times 10^3\,\mathrm{kg\,m^{-3}}$, $\gamma = 36.3\,\mathrm{mN\,m^{-1}}$). In microspace with a d_h of 100 μm, the interfacial tension exceed gravity by three orders of magnitude.

Re is defined as

$$
\begin{aligned}
Re &= \frac{\rho u d_h}{\mu} \\
&= \frac{2\rho u w}{\mu(w + 2d)}
\end{aligned} \tag{4}
$$

where ρ, u, and μ are the density, the mean velocity, and the viscosity, respectively. When water flows in the microchannel at the flow rate of $1\,\mathrm{\mu l\,min^{-1}}$, Re becomes only 0.05 in microspace with a d_h of 100 μm. Therefore, multiphase microflows are considered as laminar flow.

When physical and properties of the multiphase microflows such as viscosity and interfacial tension can be regarded as constant, the flow can be divided in two parts. One is a transient part where the flow profile is gradually approaching an equilibrium state after confluence, and the other is an equilibrium part. The flow vector profile of the latter can be expressed analytically with simple assumptions. At first, two-phase microflows in the equilibrium state is considered. The two-phase microflows are illustrated in Figure 11. Phase I and II have viscosities of μ^I and μ^{II} and widths of a and b, respectively. As driving force of the flow, pressure difference of $P_L - P_0 \equiv \Delta P$ is assumed for a channel length of L. The x- and z-axes are defined as directions across and along the channel, respectively. The origin of the x-axis is assumed at the interface of phases I and II. Under these conditions, shear stress τ_{xz} can be expressed based on momentum balance as

$$\frac{d\tau_{xz}}{dx} = \frac{\Delta P}{L} \tag{5}$$

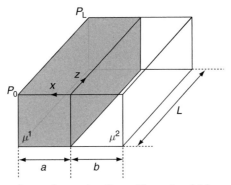

Figure 11   Illustration of two-phase microflows. Phases I and II have viscosities of μ^I and μ^{II}, and widths of a and b, respectively.

In order to obtain the stress for phases I and II, integral calculus of Equation (4) is expressed as

$$\tau^I_{xz} = \frac{\Delta P}{L}x + C^I_1$$
$$\tau^{II}_{xz} = \frac{\Delta P}{L}x + C^{II}_1 \tag{6}$$

where C_1 is constant and superscripts I and II mean phases I and II. Here, continuity of shear stress is assumed as a boundary condition. Namely,

$$\tau^I_{xz} = \tau^{II}_{xz} \text{ at } x = 0 \tag{7}$$

By substituting Equation (6) with Equation (7), the following relationship is obtained.

$$C^I_1 = C^{II}_1 = C_1 \tag{8}$$

Here, Newton's law of viscosity is used,

$$\tau_{xz} = \mu \frac{dv_z}{dx}\tau^{II}_{xz} \tag{9}$$

where v_z is flow velocity in the z-direction. By using Equations (6)–(9), the velocities of phases I and II, v^I and v^{II}, are expressed as

$$v^I_z = \frac{\Delta P}{2L\mu^I}x^2 + \frac{C_1}{\mu^I}x + C^I_2$$
$$v^{II}_z = \frac{\Delta P}{2L\mu^{II}}x^2 + \frac{C_1}{\mu^{II}}x + C^{II}_2 \tag{10}$$

where C_2 is a constant. Here, consistency of v_z^{I} and v_z^{II} at the interface and nonslip conditions are assumed as boundary conditions. Namely,

$$
\begin{aligned}
v_z^{\text{I}} &= v_z^{\text{II}} \text{ at } x = 0 \\
v_z^{\text{I}} &= 0 \text{ at } x = a \\
v_z^{\text{II}} &= 0 \text{ at } x = -b
\end{aligned}
\tag{11}
$$

From Equations (10) and (11), C_1, C_2^{I}, and C_2^{II} are eliminated as

$$
v_z^{\text{I}} = \frac{\Delta P}{L} \frac{a^2}{\mu^{\text{I}}} \left[1 + \frac{\frac{b^2}{a^2}\mu^{\text{I}} - \mu^{\text{II}}}{\frac{b}{a}\mu^{\text{I}} + \mu^{\text{II}}} + \frac{\frac{b^2}{a^2}\mu^{\text{I}} - \mu^{\text{II}}}{\frac{b}{a}\mu^{\text{I}} + \mu^{\text{II}}}\left(\frac{x}{a}\right) - \left(\frac{x}{a}\right)^2 \right]
\tag{12}
$$

$$
v_z^{\text{II}} = \frac{\Delta P}{L} \frac{b^2}{\mu^{\text{II}}} \left[1 - \frac{\mu^{\text{I}} - \frac{a^2}{b^2}\mu^{\text{II}}}{\mu^{\text{I}} + \frac{b}{a}\mu^{\text{II}}} + \frac{\mu^{\text{I}} - \frac{a^2}{b^2}\mu^{\text{II}}}{\mu^{\text{I}} + \frac{b}{a}\mu^{\text{II}}}\left(\frac{x}{b}\right) - \left(\frac{x}{b}\right)^2 \right]
\tag{13}
$$

From Equations (12) and (13), the flow velocity profile can be calculated. By integrating these equations along the x-direction, average velocities for phases I and II, $\langle v_z^{\text{I}} \rangle$ and $\langle v_z^{\text{II}} \rangle$, are expressed as

$$
\langle v_z^{\text{I}} \rangle = \frac{1}{a}\int_0^a v_z^{\text{I}}\, dx = \frac{\Delta P}{12L} \frac{a^2}{\mu^{\text{I}}} \left(4 + \frac{3}{a}\frac{b^2\mu^{\text{I}} - a^2\mu^{\text{II}}}{b\mu^{\text{I}} + a\mu^{\text{II}}} \right)
\tag{14}
$$

$$
\langle v_z^{\text{II}} \rangle = \frac{1}{b}\int_{-b}^0 v_z^{\text{II}}\, dx = \frac{\Delta P}{12L} \frac{a^2}{\mu^{\text{II}}} \left(4 - \frac{3}{b}\frac{b^2\mu^{\text{I}} - a^2\mu^{\text{II}}}{b\mu^{\text{I}} + a\mu^{\text{II}}} \right)
\tag{15}
$$

The volume flow ratio to achieve the width ratio of a/b is expressed as

$$
\frac{b\langle v_z^{\text{II}} \rangle}{a\langle v_z^{\text{I}} \rangle} = \frac{\mu^{\text{I}}}{\mu^{\text{II}}} \frac{b^3}{a^3} \frac{b^2\mu^{\text{I}} + (4ab + 3a^2)\mu^{\text{II}}}{(3b^2 + 4ab)\mu^{\text{I}} + a^2\mu^{\text{II}}} = \frac{b}{a} \frac{3 + 4\left(\frac{b}{a}\right) + \left(\frac{b}{a}\right)^2 \frac{\mu^{\text{I}}}{\mu^{\text{II}}}}{3 + 4\left(\frac{a}{b}\right) + \left(\frac{a}{b}\right)^2 \frac{\mu^{\text{II}}}{\mu^{\text{I}}}}
\tag{16}
$$

When the width of the phase I is equal to that of phase II, $a = b$, the flow velocity can be expressed as

$$
v_z^{\text{I}} = \frac{\Delta P}{2L} \frac{a^2}{\mu^{\text{I}}} \left[\frac{2\mu^{\text{I}}}{\mu^{\text{I}} + \mu^{\text{II}}} + \frac{\mu^{\text{I}} - \mu^{\text{II}}}{\mu^{\text{I}} + \mu^{\text{II}}}\left(\frac{x}{a}\right) - \left(\frac{x}{a}\right)^2 \right]
\tag{17}
$$

$$
v_z^{\text{II}} = \frac{\Delta P}{2L} \frac{a^2}{\mu^{\text{II}}} \left[\frac{2\mu^{\text{II}}}{\mu^{\text{I}} + \mu^{\text{II}}} + \frac{\mu^{\text{I}} - \mu^{\text{II}}}{\mu^{\text{I}} + \mu^{\text{II}}}\left(\frac{x}{a}\right) - \left(\frac{x}{a}\right)^2 \right]
\tag{18}
$$

In the same way, average velocities for $a = b$ can be expressed as

$$\left\langle v_z^{\mathrm{I}} \right\rangle = \frac{1}{a} \int_0^a v_z^{\mathrm{I}} \, dx = \frac{\Delta P}{12L} \frac{a^2}{\mu^{\mathrm{I}}} \left(\frac{7\mu^{\mathrm{I}} + \mu^{\mathrm{II}}}{\mu^{\mathrm{I}} + \mu^{\mathrm{II}}} \right) \tag{19}$$

$$\left\langle v_z^{\mathrm{II}} \right\rangle = \frac{1}{b} \int_{-b}^0 v_z^{\mathrm{II}} \, dx = \frac{\Delta P}{12L} \frac{a^2}{\mu^{\mathrm{II}}} \left(\frac{\mu^{\mathrm{I}} + 7\mu^{\mathrm{II}}}{\mu^{\mathrm{I}} + \mu^{\mathrm{II}}} \right) \tag{20}$$

The volume flow ratio to achieve the equal widths is expressed as

$$\frac{b \left\langle v_z^{\mathrm{II}} \right\rangle}{a \left\langle v_z^{\mathrm{I}} \right\rangle} = \frac{\mu^{\mathrm{I}}}{\mu^{\mathrm{II}}} \frac{\mu^{\mathrm{I}} + 7\mu^{\mathrm{II}}}{7\mu^{\mathrm{I}} + \mu^{\mathrm{II}}} = \frac{7 + \dfrac{\mu^{\mathrm{I}}}{\mu^{\mathrm{II}}}}{7 + \dfrac{\mu^{\mathrm{II}}}{\mu^{\mathrm{I}}}} \tag{21}$$

Figure 12 shows flow velocity profiles calculated based on Equations (17) and (18) for the two-phase microflows of water (1.0 cP)/octanol (7.3 cP),

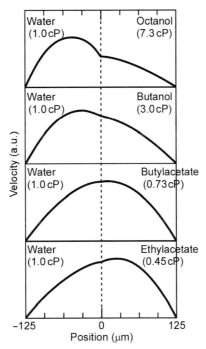

Figure 12 Calculated flow velocity profiles for 4 two-phase microflows with the same pressure. Channel width of 250 μm is assumed.

water/butanol (3.0 cP), water/butyl acetate (0.73 cP), and water/ethyl acetate (0.43 cP).

Multiphase microflows are dominated by pressures (Aota et al., 2007a, 2009a). One important parameter needed to describe the multiphase microflows is the pressure that drives the fluids. The pressure decreases in the downstream part of the flow because of the fluids' viscosity. When two fluids in contact with one another have different viscosities, the pressure difference (ΔP_{Flow}) between the two phases is a function of the contact length and the flow velocity. Another important parameter is the Laplace pressure ($\Delta P_{\text{Laplace}}$) caused by the interfacial tension between two phases. The position of the interface is fixed within a point in the microchannel by the balance established between the $\Delta P_{\text{Laplace}}$ and ΔP_{Flow}.

Figure 13a illustrates the pressure balance at the liquid–liquid interface of the two-phase microflows. The liquid–liquid interface curves toward the organic phase in a glass surface because of the hydrophilicity of the glass. $\Delta P_{\text{Laplace}}$ is generated at the curved liquid–liquid

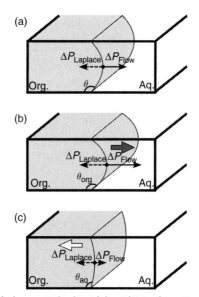

Figure 13 (a) Pressure balance at the liquid–liquid interface. Pressure difference is balanced with Laplace pressure when the two phases are separated. (b) When the pressure difference between the two phases is larger than the maximum Laplace pressure, the organic phase moves toward the aqueous phase. (c) When the pressure difference is lower than the minimum Laplace pressure, the aqueous phase moves toward the organic phase. Abbreviation: Aq., aqueous phase, Org., organic phase (Aota et al., 2009a).

interface. On the basis of the Young–Laplace equation, $\Delta P_{Laplace}$ is estimated as follows:

$$\Delta P_{Laplace} = \gamma\left(\frac{1}{R_1} + \frac{1}{R_2}\right) \tag{22}$$

where R_1 and R_2 are the curvature radii of the liquid–liquid interface in direction vertical and parallel to the liquid stream. For parallel multiphase microflows, the equation is simplified as follows:

$$\Delta P_{Laplace} = \gamma\left(\frac{1}{R_1}\right) = \frac{2\gamma \sin(\theta - 90°)}{d} \tag{23}$$

where θ and d are the contact angle and depth of the microchannels. The contact angle is restricted to the values between the advancing contact angle of the aqueous phase θ_{aq} and that of the organic phase θ_{org}. Therefore, $\Delta P_{Laplace}$ is restricted as follows:

$$\frac{2\gamma \sin(\theta_{aq} - 90°)}{d} < \Delta P_{Laplace} < \frac{2\gamma \sin(\theta_{org} - 90°)}{d} \tag{24}$$

When ΔP_{Flow} exceeds the maximum $\Delta P_{Laplace}$, the organic phase flows toward the aqueous phase (Figure 13b). When ΔP_{Flow} is lower than the minimum $\Delta P_{Laplace}$, the aqueous phase flows toward the organic phase (Figure 13c). When the flow rate ratio is changed, the pressure balance is maintained by changing the position of the liquid–liquid interface. This model indicates that the important parameters for microfluid control are the interfacial tension, the dynamic contact angle, and the depth of the microchannel. This model can also be applied to gas–liquid microflows.

Considering the interfacial pressure model, the flow rates should be between the higher and lower limits. ΔP_{Flow} can be evaluated by considering the pressure loss in the microchannels. Assuming that the pressure at the opened outlet is atmospheric pressure, P_{atm}, then the pressure P of each phase from the pressure loss ΔP is expressed as follows:

$$\begin{aligned} P &= P_{atm} + \Delta P_{tube} + \Delta P_{channel} \\ &= P_{atm} + \frac{2f\,\rho v_{tube}^2 L_{tube}}{d_{h,\,tube}} + \frac{2f\,\rho v_{channel}^2 L_{channel}}{d_{h,\,channel}} \end{aligned} \tag{25}$$

where f, ρ, v, and L are the friction factor, the density, the mean velocity of the fluid, and the length of the tube and channel. The subscripts of tube and channel correspond to the parts of the outlet tube and the microchannel. When the gas–liquid and liquid–liquid microflows are considered to be laminar flow, f is expressed as follows:

$$f = \frac{16}{Re} = \frac{16\mu}{\rho v d_h} \tag{26}$$

Therefore, ΔP_{Flow} can be expressed as follows:

$$\Delta P_{Flow} = P_{org} - P_{aq}$$

$$= \frac{32\mu_{org}v_{channel,org}L_{channel,org}}{d_{h,\,channel,org}{}^2} + \frac{32\mu_{org}v_{tube,org}L_{tube,org}}{d_{h,\,tube,org}{}^2}$$

$$- \frac{32\mu_{aq}v_{channel,aq}L_{channel,aq}}{d_{h,\,channel,aq}{}^2} - \frac{32\mu_{org}v_{tube,aq}L_{tube,aq}}{d_{h,\,tube,aq}{}^2} \qquad (27)$$

where the subscripts org and aq correspond to the organic and aqueous phases, respectively. By utilizing Equation (27), the higher and lower limits of ΔP_{Flow} were evaluated. Table 2 summarizes the literature values of the viscosities of various solvents. $\Delta P_{Laplace}$ can be evaluated by utilizing Equation (23). Table 3 summarizes the interfacial tension of various solvents. The advancing and receding contact angles of water on a glass

Table 2   Summary of viscosity of solvent at 20 °C

Solvent	Viscosity (cP)
Acetone	0.315
Aniline	4.40
1-Butanol	2.948
Butyl acetate	0.732
Carbon tetrachloride	0.969
Chloroform	0.58
Cyclohexane (at 17 °C)	1.02
1-Decanol	12.96
Dichloromethane	0.445
Diethyl ether (at 25 °C)	0.224
1,4-dioxane (at 25 °C)	1.177
Dodecane (at 25 °C)	1.383
Ethanol	1.20
Ethyl acetate	0.455
Hexane	0.326
Methanol	0.597
Nitrobenzene	2.03
1-Octanol (at 25 °C)	7.288
Pentane	0.240
1-Propanol	2.256
Toluene	0.590
Water	1.002
m-Xylene	0.620

Table 3 Summary of interfacial tension between air–solvent and aqueous–organic interfaces at 20 °C

Solvent	For air (mN m^{-1})	For water (mN m^{-1})
Acetone	25.3	
Aniline		6.1
1-Butanol		1.6
Butyl acetate		13.5
Carbon tetrachloride	27.6	
Chloroform	28.7	30.8
Cyclohexane		47.7
1-Decanol	28.6	
Dichloromethane	31.1	28.4
Diethyl ether	19.1	
1,4-Dioxane	34.2	
Dodecane	25.3	
Ethanol	22.7	
Ethyl acetate	24.3	6.3
Hexane	19.4	50.0
Methanol	23.5	
Nitrobenzene	43.3	25.2
1-Octanol	27.7	
Pentane	18.5	
1-Propanol	23.9	
Toluene	28.8	35.3
Water	72.0	
m-Xylene	28.7	36.6

plate in toluene are 10.2 ± 4.9° and 64.5 ± 4.3°, respectively. These values were also used for the contact angles in other organic solvents.

The phase separation conditions of liquid–liquid microflows for various organic solvents in the double-Y-type microchannel with a width of 215 μm, a depth of 34 μm, and a contact length of 20 mm are shown in Figure 14. The flow rate of the aqueous phase was fixed to be 10 μl min^{-1}. The experimental results agreed well with the theoretical values. The experimental results also agreed with simulated values obtained by the three-dimensional simulation using the volume of fluid method. Controlling ΔP_{Flow} and ΔP_{Laplce} permits phase separation to be achieved in the microchannel. With the help of this, the researchers can design various microfluidic chemical processes using liquid–liquid microflows.

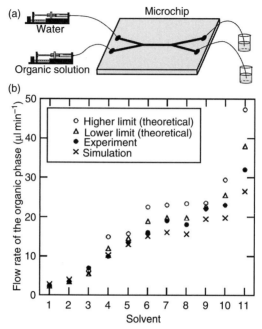

Figure 14 (a) Illustration of the microchip having a width of 215 μm, a depth of 35 μm, and a contact length of 20 mm. (b) Phase separation conditions of the liquid–liquid microflows. Solvents: 1, aniline; 2, 1-butanol; 3, nitrobenzene; 4, cyclohexane; 5, butyl acetate; 6, *m*-xylene; 7, chloroform; 8, toluene; 9, ethyl acetate; 10, dichloromethane; and 11, hexane. The opened circles show the theoretical higher limit, the opened triangles the theoretical lower limit, the solid circles experimental results, and the crosses the results of the simulation (Aota et al., 2009a).

The phase separation conditions of air–liquid microflows for various solvents in the double-Y-type microchannel with a width of 100 μm, a depth of 45 μm, and a contact length of 20 mm are shown in Figure 15. The flow rate of the air phase was fixed to be 1 ml min^{-1}. The results, however, did not agree with the theoretical values. Assuming only the pressure loss and the Laplace pressure, the phase separation should be achieved within the narrow range of the flow rates. However, phase separation was achieved over a wide range of flow rates. This disparity may be explained by considering the compression of the gas phase, evaporation of the liquid phase, and wetting at the outlet port of the liquid phase. Therefore, when gas–liquid microflows form in microchannels, the compressivity, vapor pressure, humidity, and pressure of the outlet should be carefully considered.

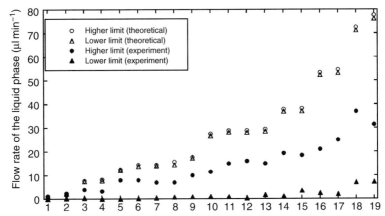

Figure 15   Phase separation conditions of the gas–liquid microflows. Solvents: 1, 1-decanol; 2, 1-octanol; 3, 1-propanol; 4, nitrobenzene; 5, dodecane; 6, 1,4-dioxane; 7, ethanol; 8, water; 9, carbon tetrachloride; 10, *m*-xylene; 11, hexane; 12, toluene; 13, chloroform; 14, ethyl acetate; 15, dichloromethane; 16, hexane; 17, acetone; 18, pentane; and 19, diethyl ether. The open circles show the theoretical higher limit, the open triangles the theoretical lower limit, the solid circles the experimental results of the higher limit, and the solid triangles the experimental results of the lower limit (Aota et al., 2009a).

3.2 Methods of stabilization of multiphase microflows

Disturbances in the flow rates of pumps can destabilize multiphase microflows. However, the control method utilizing the microchannel structure and surface energy is effective for maintaining robust control over microfluidics. Researchers have suggested that multiphase microflows can be stabilized by altering the structure or adding features to the microchannels, specifically by including a guide structure (Tokeshi et al., 2002) or a pillar structure (Maruyama et al., 2004) along the microchannel, which permit a wider range of contact angle for the fluid interface than a flat surface. Researchers can design various microfluidic chemical processes using liquid–liquid microflows.

Other groups have proposed selective chemical surface modification for stabilization of the multiphase microflows (Aota et al., 2007a, 2007b; Hibara et al., 2002, 2003, 2005, 2008; van der Linden et al., 2006; Zhao et al., 2001, 2002a, 2002b, 2003) Figure 16 illustrates the shape of the liquid–liquid interface in a microchannel with chemically patterned surfaces. The contact angle of water on a hydrophobic surface can be larger than 90°. Therefore, multiphase microflows in microchannels with patterned surfaces can form under a wider range of conditions compared to microchannels with a guide structure or pillar structure.

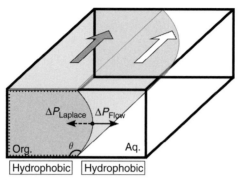

Figure 16 Shape of the liquid–liquid interface, whose contact line is pinned at the boundary between the hydrophilic and the hydrophobic surfaces.

Beebe's group has proposed a selective chemical surface modification method by combining multiphase laminar microflow and self-assembled monolayer (SAM) chemistry (Zhao et al., 2001, 2002a, 2002b, 2003). The flow in the microchannels is laminar flow as described in Section 3.1. Therefore, multiphase microflows of miscible solutions can flow side-by-side without turbulent mixing. A stream of pure hexadecane and a stream of octadecyltrichlorosilane (ODS) solution in hexadecane were introduced together in microchannels by syringe pumps, and laminar flow was maintained for 2–3 min (Figure 17). SAMs formed on both the top and bottom of the microchannels in the area containing the ODS solution. The other microchannels in the area without the ODS solution remain the hydrophilicity of the bare glass surface. The flow rates must be fast in order to prevent diffusion of ODS.

Beebe's group has also proposed a selective chemical surface modification method by photolithography (Zhao et al., 2001, 2002a, 2002b, 2003). This method uses photocleavable SAMs having hydrophobic and hydrophilic groups. First, the microchannels were modified using the photocleavable SAMs. The microchannels were cleaned by sequentially flushing

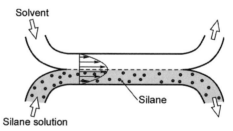

Figure 17 Schematic illustration of the selective surface modification using multiphase laminar microflows (Zhao et al., 2001).

with hexane and methanol after monolayer deposition from a 0.5 wt % solution of the corresponding trichlorosilane in hexadecane, and then dried with nitrogen. A photomask is placed on top of the SAM-modified microchannels filled with NaOH solution. Ultraviolet (UV) irradiation through a mask for 90 min leads to the production of hydrophilic groups in the irradiated regions (Figure 18).

Kitamori's group has proposed selective chemical surface modification utilizing capillarity (called the capillarity restricted modification or CARM method) (Hibara et al., 2005). In the CARM method, a microchannel structure combining shallow and deep microchannels and the principle of capillarity are utilized. The procedures are shown in Figure 19. A portion of an ODS/toluene solution (1 wt%) is dropped onto the inlet hole of the shallow channel, and the solution is spontaneously drawn into this channel by capillary action. The solution is stopped at the boundary between the shallow and deep channels by the balance between the solid–liquid and gas–liquid interfacial energies. Therefore, the solution does not enter the deep channel. It remains at the boundary for several minutes and is then pushed from the deep channel side by air pressure.

Since $\Delta P_{\mathrm{Laplace}}$ depends on the depth of the microchannels, more effective stabilization of multiphase microflows can be attained by utilizing shallow microchannels. For example, a microstructure combining shallow and deep microchannels with hydrophobic and hydrophilic surfaces, respectively, can yield large $\Delta P_{\mathrm{Laplace}}$. In a shallow 10-μm-deep microchannel, $\Delta P_{\mathrm{Laplace}}$ of the gas–water interface is 6 kPa. Microchannels with an asymmetric cross section with the patterned surfaces mentioned above are effective not only for the stability of the multiphase microflows, but also for a fail-safe system required for practical systems. Air bubbles contamination can disturb the stability of systems and liquid–liquid two-phase microflows sometimes are unstable due to some disturbances. By using microchannels having asymmetric cross sections along with the patterned surface, one can convert plug flow into two-phase microflows (Figure 20) (Hibara et al., 2005, 2008).

In chemically patterned microchannels, microcountercurrent flows can form (Aota et al., 2007a, 2007b, 2007c). In conventional macroscale devices, countercurrent flow is attained by gravitational segregation involving droplets (Figure 21a). However, parallel countercurrent flow in the laminar flow regime cannot be easily realized. In an ordinary microchannel, countercurrent flow cannot form because high shear stress at the interface causes breakup (Figure 21b) and the two phases collide (Figure 21c). To form parallel microcountercurrent flows, the aqueous solution must flow along one side of the channel and the organic solution must flow along the other side without breakup. Considering pressure balance at the interface, microcountercurrent flows can be formed in a microchannel with patterned surfaces. The phase separation conditions for gas–liquid and

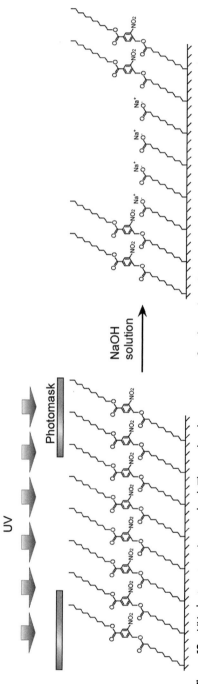

Figure 18 UV photopatterning method. The molecular structure of a photocleavable SAM formed on glass surfaces. UV irradiation through masks placed on top of SAM-modified microchannels leads to the production of hydrophilic carboxylate groups in the irradiated regions (Zhao et al., 2001).

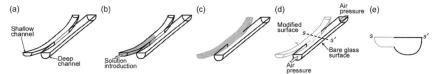

Figure 19 Modification procedures by CARM method. (a) The shallow and deep microchannels have separate inlet holes and contact points in the microchip. (b) A solution containing modification compounds is introduced from the inlet of the shallow microchannel by capillarity. (c) The solution does not leak to the deep microchannel and only the shallow microchannel is modified. (d) The solution is pushed away with air pressure from the deep microchannel. (e) A sectional illustration along the s–s' dashed line in (d) (Hibara et al., 2005).

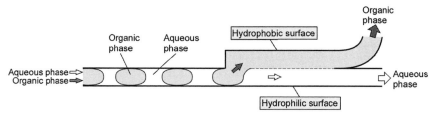

Figure 20 Conversion of plug flow into parallel two-phase microflows (Hibara et al., 2008).

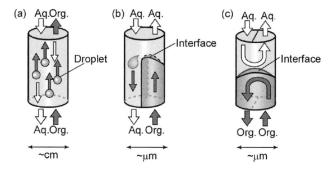

Figure 21 Schematic illustration of (a) countercurrent flows in conventional macroscale devices, (b) droplet generation because of breakup due to high shear stress in an ordinary microchannel, and (c) collision of two phases in an ordinary microchannel. Abbreviation: Aq., aqueous phase, Org., organic phase (Aota et al., 2007c).

liquid–liquid microcountercurrent flows in the microchannel with an asymmetric cross section are shown in Figure 22. Since the viscosity of air can be thought of as being negligible, the maximum flow rate of water becomes constant under experimental conditions. The theoretical higher limit value is shown as the solid line in Figure 22b. Experimental results

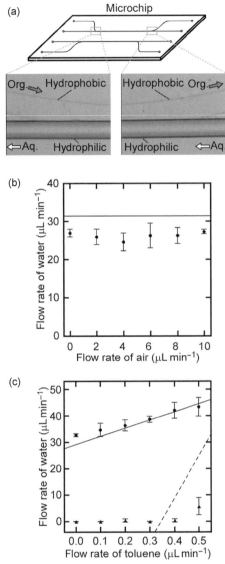

Figure 22 (a) Optical microscope images of the phase separation at the confluences. (b) Maximum flow rate of water as a function of the flow rate of air. The solid circles show the experimental maximum flow rates and the solid line theoretical higher limit. (c) Maximum and flow rates of water as a function of the flow rate of toluene. The solid circles show the experimental maximum flow rates, the solid triangles the experimental minimum flow rates, the solid line theoretical higher limit, and the dashed line the theoretical lower limit (Aota et al., 2007a).

differed slightly from the theoretical value, a discrepancy that could be explained by experimental issues around the outlet of the microchannel, such as pressure loss and wetting. Considering the theoretical lower limit, the viscosity of air of 18.6 µPa s at 300 K is only 1.9% that of water. Therefore, the pressure of the air phase can be considered to be negligible. Figure 22c shows the phase separation conditions of the aqueous toluene microcountercurrent flows. For liquid–liquid microcountercurrent flows, the viscosities of both phases play an important role. Therefore, the higher and lower limits are not constant. The theoretical higher and lower limit values are shown as the solid and dashed lines in Figure 22c, respectively. The maximum flow rate in the upstream part of the aqueous phase agreed well with the theoretical value. The minimum flow rate, however, did not agree with the theoretical values. This disparity could be explained in terms of the microchannel geometry. When water leaks onto the hydrophobic surface, the contact angle on the upper wall is the same as that on the lower wall because the hydrophobic channel is flat. However, when toluene leaks onto the hydrophilic surface, the contact angle on the upper wall is different from that on the lower wall because the hydrophilic channels has a semicircular cross section and the boundary between the hydrophilic and hydrophobic surfaces has an edge. Since the range of contact angles of the liquid–liquid interface at the edge becomes wider compared to that on a flat surface, the liquid–liquid interface at the edge structure can produce a larger value of $\Delta P_{Laplace}$ than from a flat surface. Since the edge structure can expand the range of flow rate conditions for phase separation, this disagreement may not be a serious problem for the design of microfluidic chemical processes utilizing multiphase microflows.

Gas–liquid and liquid–liquid microflows are effective for highly efficient extraction processes. When samples comprise adsorbent molecules at a fluid–liquid interface, molecular adsorption changes the interfacial tension of the fluid–liquid interface. Since the specific interfacial area is very large in a microchip, the change in interfacial tension due to molecular adsorption can be an important factor affecting phase separation of the gas–liquid and liquid–liquid microflows. The molecular concentration is higher in the downstream portion of the microflows. Thus, when designing a chemical process that includes molecular transport through the interface, researchers need to modify the proposed model by considering molecular adsorption at the interface. For microextraction design, the *in situ* interfacial tension measurement is especially important. The microscopic quasielastic laser scattering method could be a powerful tool for this purpose (Hibara et al., 2003).

Superhydrophilic and superhydrophobic surfaces are more effective at stabilizing two-phase microflows. These surfaces can be obtained by creating roughness utilizing titanium nanoparticles. Titanium modification of a microchannel yields nanometer-scale surface roughness, and subsequent hydrophobic treatment creates a superhydrophobic surface. Photocatalytic decomposition of the coated hydrophobic molecules was used to pattern the surface wettability, which was tuned from superhydrophobic to super-hydrophilic under controlled photoirradiation (Takei et al., 2007a). This method can also be applied to the conversion of plug flow into two-phase microflows.

3.3 Wettability-based microvalve

Numerous miniaturized mechanical valves fabricated by micromachining technologies have been developed (Auroux et al., 2002; Dittrich et al., 2006; Reyes et al., 2002). Pneumatic-controlled valves made of soft material have also been reported (Unger et al., 2000). All of these valves require both fabrication and construction of mechanical moving parts or the pressure of external control parts in the microchannels. Also, there have been reports of nonmechanical methods utilizing phase transition solution (Gui and Liu, 2004), a thermoresponsive polymer (Yu et al., 2003), or hydrogel (Beebe et al., 2000). Additionally, patterned hydrophobic surfaces have been demonstrated as valves for microfluid control in a variety of applications. The patterned titanium surfaces described previously can be used very effectively as valves in a microchip due to their superhydrophobic nature (Takei et al., 2007b). Batch operation in the microchannels with patterned surfaces has been demonstrated as shown in Figure 23. First, fluorescent solution was introduced at a pressure below the maximum Laplace pressure (Figure 23a). Second, air was introduced at the same pressure to push out excess solution and retain the plug of picoliter volume (Figure 23b). Third, air was introduced at a pressure

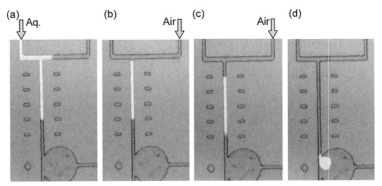

Figure 23 Fluorescence images of the liquid motion during batch operation in a microchannel. The liquid is (a) being introduced, (b) measured, (c) transferred, and (d) dispensed to the other channel (Takei et al., 2007b).

higher than the maximum Laplace pressure to expel the plug (Figure 23c, d). This microvalve-utilizing chemical surface modification has no dead volume and can supply a constant volume of liquid for reaction or analysis. Moreover, the repeatability of this operation allows it to be applied to titration.

4. DISCUSSION AND CONCLUSIONS

The basic methodologies and applications for microfluidics were introduced. In Chapter 2, the general concepts (MUO and CFCP) for microintegration and applications based on CFCP will be discussed. CFCP approach can be applied to develop more complicated chemical processing system. If CAD for microchemical processing, which includes microfluid simulation, reaction time analysis, extraction time analysis, and microchennel structure design, the time of development for micro-chemical systems would be drastically shortened. In Chapter 3, fluid control methods will be discussed. The viscous force and the interfacial tension are effective forces in microspaces. In order to control these forces, surface chemistry and microchannel structures play important roles. Since the basic methodologies have already been developed, devel-opment of microfluidic systems for practical applications is required in the next stage.

LIST OF SYMBOLS

a	width of phase I (m)
b	width of phase II (m)
d	depth (m)
d_h	equivalent diameter (m)
f	friction factor (-)
g	gravitational acceleration ($m\,s^{-2}$)
K	diffusion coefficient ($m^2\,s^{-1}$)
L	channel length (m)
l_p	perimeter (m)
l_d	diffusion length (m)
P	pressure ($N\,m^{-2}$)
P_{atm}	atmospheric pressure ($N\,m^{-2}$)
R	curvature radius (m)
S	cross section (m^2)
t	diffusion time (s)
u	mean velocity ($m\,s^{-1}$)

| v | flow velocity (m/s) |
| w | width (m) |

GREEK LETTERS

γ	interfacial tension (N m^{-1})
μ	viscosity (cP)
ΔP	pressure difference (N m^{-2})
ΔP_{Flow}	pressure difference between two phases (N m^{-2})
ΔP_{Laplce}	Laplace pressure (N m^{-2})
$\Delta\rho$	density difference (kg m^{-3})
θ	contact angle (degree)
θ_{aq}	advancing contact angle of aqueous phase (degree)
θ_{org}	advancing contact angle of organic phase (degree)
ρ	density (kg m^{-3})
τ	shear stress (N m^{-2})

REFERENCES

Aota, A., Hibara, A., and Kitamori, T. *Anal. Chem.* **79**, 3919–3924 (2007a).

Aota, A., Hibara, A., and Shinohara, K., et al. *Anal. Sci.* **23**, 131–133 (2007b).

Aota, A., Mawatari, K., and Takahashi, S., et al. *Microchim. Acta* **164**, 249–255 (2009a).

Aota, A., Mawatari, K., and Kihira, Y., et al. Micro continuous gas analysis system of ammonia in cleanroom, *in* "Proceedings of Micro Total Analysis Systems 2009" (T. Kim, Y. S. Lee, T. D. Chung, N. L. Jeon, S. H. Lee, K. Y. Suh, J. Choo, Y. K. Kim Eds.), pp. 609–611. The Chemical and Biological Microsystems Society, Jeju-island, Korea (2009b).

Aota, A., Nonaka, M., and Hibara, A., et al. *Angew. Chem. Int. Ed.* **46**, 878–880 (2007c).

Auroux, P. A., Lossifidis, D., and Reyes, D. R., et al. *Anal. Chem.* **74**, 2637–2652 (2002).

Beebe, D. J., Moore, J. S., and Bauer, J. M., et al. *Nature* **404**, 588–590 (2000).

Brody, J. P., and Yager, P. *Sens. Actuators, A* **58**, 13–18 (1997).

Burns, J. R., and Ramshaw, C. *Lab Chip* **1**, 10–15 (2001).

Dittrich, P. S., Tachikawa, K., and Manz, A. *Anal. Chem.* **78**, 3887–3908 (2006).

Goto, M., Sato, K., and Murakami, A., et al. *Anal. Chem.* **77**, 2125–2131 (2005).

Goto, M., Tsukahara, T., and Sato, K., et al. *Anal. Bioanal. Chem.* **390**, 817–823 (2008).

Gui, L., and Liu, J. *J. Micromech. Microeng.* **14**, 242–246, (2004).

Hachiya, H., Matsumoto, T., and Kanda, K., et al., Micro environmental gas analysis system by using gas-liquid two phase flow, *in* "Proceedings of Micro Total Analysis Systems 2004" (T. Laurell, J. Nillson, K. F. Jensen, D. J. Harrison, J. P. Kutter Eds.), pp. 91–101. Royal Society of Chemistry, Mölme, Sweden (2004).

Hibara, A., Iwayama, S., and Matsuoka, S., et al. *Anal. Chem.* **77**, 943–947 (2005).

Hibara, A., Kasai, K., and Miyaguchi, H., et al. Novel two-phase flow control concept and multi-step extraction microchip, *in* "Proceedings of Micro Total Analysis Systems 2008" (L. E. Locascio, M. Gaitan, B. M. Paegel, D. J. Ross, W. N. Vreeland Eds.), pp. 1326–1328. Chemical and Biological Microsystems Society, San Diego, USA, (2008).

Hibara, A., Nonaka, M., and Hisamoto, H., et al. *Anal. Chem.* **74**, 1724–1728 (2002).

Hibara, A., Nonaka, M., and Tokeshi, M., et al. *J. Am. Chem. Soc.* **125**, 14954–14955 (2003).
Hibara, A., Tokeshi, M., and Uchiyama, K., et al. *Anal. Sci.* **17**, 89–93 (2001).
Hiki, S., Mawatari, K., and Hibara, A., et al. *Anal. Chem.* **78**, 2859–2863 (2006).
Hisamoto, H., Horiuchi, T., and Tokeshi, M., et al. *Anal. Chem.* **73**, 1382–1386 (2001a).
Hisamoto, H., Horiuchi, T., and Uchiyama, K., et al. *Anal. Chem.* **73**, 5551–5556 (2001b).
Hisamoto, H., Shimizu, Y., and Uchiyama, K., et al. *Anal. Chem.* **75**, 350–354 (2003).
Hotokezaka, H., Tokeshi, M., and Harada, M., et al. *Prog. Nucl. Energy* **47**, 439–447 (2005).
Kakuta, M., Takahashi, H., and Kazuno, S., et al. *Meas. Sci. Technol.* **17**, 3189–3194 (2006).
Kikutani, Y., Hisamoto, H., and Tokeshi, M., et al. *Lab Chip* **4**, 328–332 (2004).
Kopp, M. U., de Mello, A. J., and Manz, A. *Science* **280**, 1046–1048 (1998).
Kralj, J. G., Schmidt, M. A., and Jensen, K. F. *Lab Chip* **5**, 531–535, (2005).
Maruyama, T., Kaji, T., and Ohkawa, T., et al. *Analyst* **129**, 1008–1013 (2004).
Mawatari, K., Tokeshi, M., and Kitamori, T. *Anal. Sci.* **22**, 781–784 (2006).
Minagawa, T., Tokeshi, M., and Kitamori, T. *Lab Chip* **1**, 72–75 (2001).
Miyaguchi, H., Tokeshi, M., and Kikutani, Y., et al. *J. Chromatogr. A* **1129**, 105–110 (2006).
Ohashi, T., Mawatari, K., Sato, K., et al. Automated micro-ELISA system for allergy checker: a prototype and clinical test, *in* "Proceedings of Micro Total Analysis Systems 2006" (T. Kitamori, H. Fujita, S. Hasebe Eds.), pp. 858–860. Japan Academic Association Inc., Japan (2006).
Proskurnin, M. A., Slyadnev, M. N., and Tokeshi, M., et al. *Anal. Chim. Acta* **480**, 79–95 (2003).
Reyes, D. R., Lossifidis, D., and Auroux, P. A., et al. *Anal. Chem.* **74**, 2623–2636 (2002).
Sahoo, H. R., Kralj, J. G., and Jensen, K. F. *Angew. Chem. Int. Ed.* **46**, 5704–5708 (2007).
Sato, K., Tokeshi, M., and Kimura, H., et al. *Anal. Chem.* **73**, 1213–1218 (2001).
Sato, K., Tokeshi, M., and Kitamori, T., et al. *Anal. Sci.* **15**, 641–645 (1999).
Sato, K., Tokeshi, M., and Odake, T., et al. *Anal. Chem.* **72**, 1144–1147 (2000a).
Sato, K., Tokeshi, M., and Sawada, T., et al. *Anal. Sci.* **16**, 455–456 (2000b).
Sato, K., Yamanaka, M., and Hagino, T., et al. *Lab Chip* **4**, 570–575 (2004).
Sato, K., Yamanaka, M., and Takahashi, H., et al. *Electrophoresis* **23**, 734–739 (2003).
Slyadnev, M. N., Tanaka, Y., and Tokeshi, M., et al. *Anal. Chem.* **73**, 4037–4044 (2001).
Smirnova, A., Mawatari, K., and Hibara, A., et al. *Anal. Chim. Acta* **558**, 69–74, (2006).
Smirnova, A., Shimura, K., and Hibara, A., et al. *Anal. Sci.* **23**, 103–107 (2007).
Song, H., Tice, J. D., and Ismgailov, R. F. *Angew. Chem. Int. Ed.* **42**, 767–772 (2003).
Sorouraddin, H. M., Hibara, A., and Kitamori, T. *Fresenius' J. Anal. Chem.* **371**, 91–96 (2001).
Sorouraddin, H. M., Hibara, A., and Proskrunin, M. A., et al. *Anal. Sci.* **16**, 1033–1037 (2000).
Surmeian, M., Hibara, A., and Slyadnev, M., et al. *Anal. Lett.* **34**, 1421–1429, (2001).
Surmeian, M., Sladnev, M. N., and Hisamoto, H., et al. *Anal. Chem.* **74**, 2014–2020 (2002).
Takei, G., Aota, A., and Hibara, A., et al. Phase separation of segmented flow by the photocatalytic wettability patterning and tuning of microchannel surface, *in* "Proceedings of Micro Total Analysis Systems 2007" (J. L. Viovy, P. Tabeling, S. Descroix, L. Malaquin Eds.), pp. 1213–1215. The Chemical and Biological Microsystems Society, Paris, France (2007a).
Takei, G., Nonogi, M., and Hibara, A., et al. *Lab Chip* **7**, 596–602 (2007b).
Tamaki, E., Hibara, A., and Tokeshi, M., et al. *J. Chromatogr. A* **987**, 197–204 (2003).
Tamaki, E., Hibara, A., and Tokeshi, M., et al. *Lab Chip* **5**, 129–131 (2005).
Tamaki, E., Sato, K., and Tokeshi, M., et al. *Anal. Chem.* **74**, 1560–1564 (2002).
Tanaka, Y., Sato, K., and Yamato, M., et al. *Anal. Sci.* **20**, 411–423 (2004).
Tanaka, Y., Sato, K., and Yamato, M., et al. *J. Chromatogr. A* **1111**, 233–237 (2006).
Tanaka, Y., Slyadnev, M. N., and Hibara, A., et al. *J. Chromatogr. A* **894**, 45–51 (2000).
Tokeshi, M., Minagawa, T., and Kitamori, T. *Anal. Chem.* **72**, 1711–1714 (2000a).
Tokeshi, M., Minagawa, T., and Kitamori, T. *J. Chromatogr. A* **894**, 19–23 (2000b).
Tokeshi, M., Minagawa, T., and Uchiyama, K., et al. *Anal. Chem.* **74**, 1565–1571, (2002).
Tokeshi, M., Uchida, M., and Hibara, A., et al. *Anal. Chem.* **73**, 2112–2116 (2001).

Tokeshi, M., Yamaguchi, J., and Hattori, A., et al. *Anal. Chem.* **77**, 626–630 (2005).

Unger, M. A., Chou, H. P., and Thorsen, T., et al. *Science* **288**, 113–116 (2000).

van der Linden, H. J., Jellema, L. C., and Holwerda, M., et al. *Anal. Bioanal. Chem.* **385**, 1376–1383 (2006).

Weigl, B. H., and Yager, P. *Science* **283**, 346–347 (1999).

Yamauchi, M., Mawatari, K., and Hibara, A., et al. *Anal. Chem.* **78**, 2646–2650 (2006).

Yu, C., Mutlu, S., Selvaganapathy, P., and Mastrangelo, C. H., et al. *Anal. Chem.* **75**, 1958–1961 (2003).

Zhao, B., Moore, J. S., and Beebe, D. J. *Science* **291**, 1023–1026 (2001).

Zhao, B., Moore, J. S., and Beebe, D. J. *Anal. Chem.* **74**, 4259–4268 (2002a).

Zhao, B., Moore, J. S., and Beebe, D. J. *Langmuir* **19**, 1873–1879 (2003).

Zhao, B., Viernes, N. O.L., and Moore, J. S., et al. *J. Am. Chem. Soc.* **124**, 5284–5285 (2002b).

CHAPTER 2

Microreactors with Electrical Fields

Anıl Ağıral and **Han J.G.E. Gardeniers**[*]

Abstract   The use of electric fields in chemistry is considered an important
            concept of process intensification. The combination of electricity
            with chemistry becomes particularly valuable at smaller scales, as
            they are exploited in microreaction technology. Microreactor
            systems with integrated electrodes provide excellent platforms to

Mesoscale Chemical Systems, MESA + Institute for Nanotechnology, University of Twente, 7500 AE
Enschede, The Netherlands

[*] Corresponding author.
  E-mail: j.g.e.gardeniers@utwente.nl

Advances in Chemical Engineering, Volume 38
ISSN: 0065-2377, DOI 10.1016/S0065-2377(10)38002-1
© 2010 Elsevier Inc.
All rights reserved.

investigate and exploit electric principles as a means to control, activate, or modify chemical reactions, but also preparative separations. One example which is discussed in detail in this chapter is a microplasma, which allows chemistry at moderate temperatures with species which have a reactivity comparable to that at very high temperatures, with potential advantages in energy efficiency. Another highlighted topic is electrokinetic control of chemical reactions, which requires the small scale to operate efficiently. Electrokinetic concepts can be used to control fluid flow, but also to transport or trap particles and molecules. Several less known concepts like electric wind, electric swing adsorption, electrospray, and pulsed electric fields, are discussed, as well as examples of their application. Novel principles to control adsorption and desorption, as well as activity and orientation of adsorbed molecules are described, and the relevance of these principles for the development of new reactor concepts and new chemistry are discussed.

1. INTRODUCTION

1.1 Where chemistry meets electricity

The combination of chemistry and electricity is best known in the form of *electrochemistry*, in which chemical reactions take place in a solution in contact with electrodes that together constitute an electrical circuit. Electrochemistry involves the transfer of electrons between an electrode and the electrolyte or species in solution. It has been in use for the storage of electrical energy (in a galvanic cell or battery), the generation of electrical energy (in fuel cells), the analysis of species in solution (in pH glass electrodes or in ion-selective electrodes), or the synthesis of species from solution (in electrolysis cells).

However, there are many other options to combine electricity with chemistry. One that has been studied intensively for a variety of different applications is *plasma chemistry* (see Fridman, 2008 for a recent overview). A plasma is a partially ionized gas, in which a certain percentage of the electrons is free instead of bound to an atom or molecule. Because the charge neutrality of a plasma requires that plasma currents close on themselves in electric circuits, a plasma reactor shows resemblance to an electrochemical cell, although due to the much lower ionization degree and conductivity, a plasma discharge will typically be operated in the range of hundreds of volts, compared to a few volts in the case of an aqueous electrochemical cell.

An increasing number of industrial and commercial examples of plasma technology exist. Typical *low-pressure* discharge examples are certain types of light sources, including plasma display panels, and microfabrication processes for integrated circuit manufacturing, like sputtering, reactive ion etching and plasma-enhanced chemical vapor deposition (CVD). Arc discharges, in which a high power thermal discharge of very high temperature (~10,000 K) is generated, are typical examples of *atmospheric pressure* plasmas. Arc discharges are applied in metallurgical processing and welding. Another example of atmospheric plasma is a corona discharge, which is a non-thermal discharge generated by the application of a high voltage to sharp electrode tips. This discharge is used in ozone generators and particle precipitators.

Of particular interest is a dielectric barrier discharge (DBD). This is a nonthermal atmospheric pressure discharge generated by the application of high voltages, in which an *insulator coating* prevents the transition of the plasma discharge into an arc. The process uses a high voltage alternating at kHz to GHz frequencies. Operation at atmospheric pressure is made possible by *a small gap between the electrodes*, from 0.1 mm in plasma displays and 1 mm in ozone generators to several cm in CO_2 lasers. Thus, although DBD is applied at large scale, for example, in the surface functionalization of synthetic fabrics and plastics to improve adhesion of paints and glues (Leroux, 2006, 2008), due to the small gap required it is typically performed in *microsystems*. Furthermore, DBD, besides the possibility to work without the vacuum which is needed for most other plasma processes, operates at *low gas temperatures*, which is the reason why it has become a relevant technique for synthetic purposes. Chemistry in *microplasma devices* will be the main topic elaborated in this chapter.

The above combinations of electricity with chemistry deal with the generation of charged species in either a gas of a liquid medium. This requires ionization, which occurs by electron transfer and transport of charged species in a closed electrical circuit. The charged species themselves, or the radicals generated by them (e.g., radicals generated by electron impact in the gas phase, in a plasma) can be used as activated species taking part in chemical reactions.

On the other hand, one may also use the charges to *enhance mass transport*, either by transporting the chemical species themselves (if they carry a charge), or by transporting the medium in which the chemical species are contained. This *electrokinetic transport* exists in different forms: *electrophoresis* is the movement of charged particles in an electric field due to an electrostatic Coulomb force which drags them through the medium toward an oppositely charged surface, *electroosmosis* is the motion of a polar liquid along charged surfaces under the influence of an electric field, and *dielectrophoresis* (DEP) is based on a force exerted on a (not necessarily charged) dielectric particle subjected to a non-uniform electric field. *Electrowetting-on-dielectric*, or EWOD, can be used to induce a flow of liquid droplets on a surface or in a channel, where the droplets are pulled toward the region with high electric field. All these electrokinetic principles work best in microsystems, and will be discussed in this chapter for cases in which they are combined with synthetic chemistry.

Finally, one may use charging or polarization of *surfaces*, induced by external electric fields, to control the adsorption and desorption of molecules and the state of these adsorbed molecules, in order to control their chemical reactivity. This is an upcoming field that has not yet been explored to its fullest potential. It involves aspects of *nanotechnology* and *nanoscience*, like the fabrication of structures of several nanometers and stimuli generated by scanning tunneling microscopic probes. The outcome of the research in this field is generally of a fundamental nature. The topic of electronic control of reactions at surfaces will be discussed in the last section of this chapter.

1.2 When chemistry and electricity meet in a narrow alley

The advantages of *microreaction technology* and its establishment as a concept of process intensification have been highlighted many times before in literature, and instead of repeating that matter here, we would like to refer the reader to the many excellent reviews and monographs that have appeared over the last decade (Hessel et al., 2004, 2009; Jähnisch et al., 2004; Jensen, 2001; Kolb et al., 2007; Yoshida, 2008). In cases where microreaction technology involves electricity as a mechanism of activation or manipulation of chemical species (the use of electric fields in chemistry on its own is already considered a valuable process intensification concept, see Stankiewicz and Moulijn, 2000), the advantages of the small scale lie mainly in the shorter distances for mass transport or in the smaller gaps over which electrical fields can be applied. A typical example is, as mentioned before, a microplasma device, in which the small electrode gap allows atmospheric plasma chemistry, which is a result of the fact that although the lifetime of electrons in the gap space is limited because of the dense gas, they still live long enough to excite a significant amount of species. In addition to that, a small gap allows the use of lower voltages to obtain the same electric field strength. Other advantages of the combination of microreaction technology with electrical fields are the large surface-to-volume ratio and the possibility to integrate electrodes at relevant positions in the microreactor device.

In this chapter the focus will be on the application of electrical fields in microreactors, and the potential of such systems for chemical synthesis will be outlined. The end of the chapter will give an overview of less-studied concepts, like electronic control of surface chemistry, and will discuss the opportunities offered by nanotechnology for achieving such control.

2. MICROPLASMA REACTORS

2.1 Atmospheric pressure microplasmas

The plasma state is referred to as the fourth state of matter. It is used to describe a partially or completely ionized gas consisting of positive and negative ions, electrons, and excited and neutral species. The ionization degree of plasma can vary from partially ionized to fully ionized. Plasma exhibits quasineutrality which is referred to as a balance of positive and negative charges. Local charges can be balanced by electrostatic forces which restore the quasineutrality. Plasmas are classified by the number density (particles cm^{-3}) and average kinetic energy (eV), mostly expressed in terms of temperature (K), of the different charged species. The average kinetic energy of electrons, ions, and excited and neutral species depends on the plasma conditions. In equilibrium or thermal plasmas, all constituents (ions, electrons, and neutrals) have the same average temperature, which can vary from a few thousand Kelvin (e.g., in plasma torches) to a few million Kelvin (in fusion plasmas). In a nonthermal low-temperature plasma, the temperature of the ions and neutral species can be close to ambient temperature, while in that same plasma the temperature of the electrons can exceed several thousand Kelvin. The temperature of plasma components (e.g., electrons and species in an excited state) can exceed the temperatures applied in conventional thermal chemical processes and these exceptional conditions of the plasma can generate a thermodynamic nonequilibrium state with a high concentration of energetic and reactive species (Lieberman and Lichtenberg, 1994; Raizer, 1991; von Engel, 1955).

Atmospheric pressure nonthermal discharges may become of great importance for chemical industry because they create a highly reactive environment at cold temperatures and therewith open up alternative, highly flexible, environmentally friendly, and energy-saving processing routes. Characteristics and properties of nonequilibrium atmospheric pressure plasmas can be found in many review articles and books (Becker et al., 2004, 2005, 2006; Foest et al., 2006; Lieberman and Lichtenberg, 1994; Raizer, 1991; Tachibana, 2006; von Engel, 1955).

Atmospheric pressure plasmas, just like most other plasmas, are generated by a high electric field in a gas volume. The few free electrons which are always present in the gas, due to, for example, cosmic radiation or radioactive decay of certain isotopes, will, after a critical electric field strength has been exceeded, develop an avalanche with ionization and excitation of species. Energy gained by the hot electrons is efficiently transferred and used in the excitation and dissociation of gas molecules. In a nonequilibrium atmospheric pressure plasma, collisions and radiative processes are dominated by energy transfer by stepwise processes and three-body collisions. The dominance of these processes has allowed many novel applications, for example, in medical sterilization, biological decontamination, remediation of pollutants, excimer lamps, and light sources (Foest et al., 2006). However, high-pressure plasmas have a tendency to become unstable due to the rapid transition to arcs and filamentation. To avoid instability problems and maintain a self-sustaining discharge for practical applications, a solution was found in the confinement of the high-pressure plasma to *dimensions below about 1 mm*. Such a plasma is often referred to as a *microplasma*. The current and energy density in this type of plasma is found to be high and results in effective gas heating and momentum transfer from electrons to gas molecules.

Chemistry in reactors with dimensions below 1 mm leads us into the field of *microreactors*. Implementation of microplasmas in microreactors offers the potential to exploit the advantages of atmospheric pressure nonequilibrium chemical processes for efficient synthesis of valuable chemicals and nanostructures as well as the decomposition of hazardous compounds. As mentioned before, microplasmas can be generated at low gas temperatures and possess an electron energy distribution containing large fractions of high-energy electrons, and reactive species deriving from these electrons. In addition, the increased surface area-to-volume ratio in microreactor channels leads to enhanced plasma–surface interactions, which is very useful in cases where active coatings are present on the channel walls. The combination of the reactive species and the additional plasma activation of the surface may be exploited to produce chemical products in an energy efficient manner.

A number of configurations of microplasma reactors will be described here. Classification will be based on the power sources, the electric field switching frequency ranging from DC to GHz, and electrode geometries and materials, extending from DBDs to micro hollow cathodes and microcavity discharges.

2.1.1 DC glow discharges

Atmospheric pressure DC glow discharges can be generated between two electrodes when the current through the discharge is limited to low values by a large resistor (Grotrian, 1915). The microplasma can be stabilized when the electrode separation is kept below 1 mm and the transition to an unstable arc discharge can be avoided when the spatial dimensions of the discharge are kept small enough (Staack et al., 2005). Figure 1 shows a picture of such a discharge in air. Spectroscopic temperature measurements show that the discharge is nonthermal with a gas temperature above room temperature. The nonequilibrium nature of glow discharges for small dimensions may find applications in microreactors for gas reforming, material deposition and the destruction of environmentally harmful substances.

Figure 1 Glow discharges at atmospheric pressure in air at (a) 0.1 mm, (b) 0.5 mm, (c) 1 mm, and (d) 3 mm electrode spacing (Staack et al., 2005; reproduced with permission).

2.1.2 Dielectric barrier discharges

DBDs are nonequilibrium plasmas at atmospheric pressure with applications in ozone generation, surface modification, pollution control, excimer lamps, and recently also in flat plasma display panels (Kogelschatz, 2003). Typical planar DBD configurations are shown in Figure 2. They have at

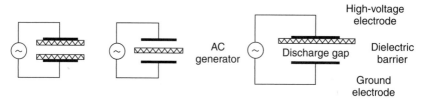

Figure 2   DBD configurations (Kogelschatz, 2003; reproduced with permission).

least one dielectric layer, being an insulator, between electrode and plasma, to prevent arc formation. Dielectric barriers can be glass, ceramic, or polymer coatings. In a DBD configuration, the plasma has a capacitive nature and consists of a large number of microdischarges in the gap between the insulator and the opposite (often uncovered) electrode, where the duration of the filamentary microdischarges is limited to a few nanoseconds (Kogelschatz, 2002). In this way, excess gas heating is minimized, although activation of molecules and atoms in the gas volume is ensured by high-energy electrons created in the microdischarges. The incorporation of a DBD in a microreactor as a miniature source of ions, excited species and radicals can generate a highly reactive and quenching environment which is difficult to obtain in thermochemical processes.

2.1.3 Micro hollow cathode discharges

Micro hollow cathode discharges (MHCDs) were first reported as stable atmospheric pressure microdischarges in cylindrical hollow cathode geometry (Schoenbach, 1996). In a typical hollow cathode structure, there is a cylindrical hole in the cathode, with a ring-shaped anode separated by an insulator (Figure 3a), or a cylindrical opening in a thin solid cathode layer (Figure 3b) (Schoenbach et al., 2003). Because of the relatively simple fabrication process of these electrode configurations, manufacturing of large area arrays of microplasma devices with parallel operation becomes feasible. Flowing gas through the plasma volume inside the hollow part allows the use of these discharges as microreactors. It is also possible to apply a third electrode placed at the anode side to achieve a stable glow discharge with dimensions of up to centimeters in

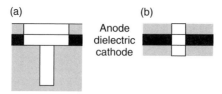

Figure 3   Electrode geometries for MHCDs (Schoenbach et al., 1996; reproduced with permission).

atmospheric pressure (Stark and Schoenbach, 1999). MHCDs can be operated at atmospheric pressure in direct current or pulsed mode with electron densities exceeding those in other nonequilibrium high-pressure glow discharges. Extreme power densities (on the order of $10^5\,\mathrm{W\,cm^{-3}}$) make these microdischarges very attractive in microreactors for the efficient decomposition of molecules, such as hydrocarbons and ammonia.

2.1.4 Microcavity discharges
Microcavity plasma devices have cavities with precisely controlled cross sections. Large arrays of these devices have been fabricated in different materials such as ceramics (Allmen et al., 2003), photodefinable glass (Kim and Eden, 2005), alumina structures (Park et al., 2005), and plastic substrates (Anderson et al., 2008). An example with inverted square pyramid microcavities fabricated in silicon is represented in Figure 4. Physical and chemical isolation between the electrodes and the discharge is maintained by the dielectric. The advantage of using silicon as the host material in these microplasma devices is the wide range of microfabrication techniques which are available for this material, which allows production of large arrays of microcavity discharge devices at reasonable expense. As an application example, parallel linear arrays of interconnected cylindrical microcavity plasma elements integrated in microreactors, based on disposable plastic substrates, have been demonstrated (Anderson et al., 2008).

2.1.5 Field emission from tip electrodes
The conditions for gas breakdown in an electrical field can be described in terms of the breakdown voltage as a function of the product of pressure and gap spacing. The resulting graph is known as the Paschen curve

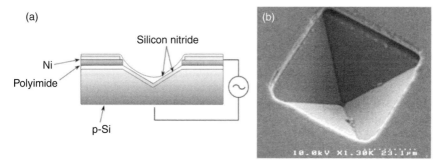

Figure 4 (a) Cross-sectional diagram of a silicon-based microcavity discharge device with an inverted square pyramid microcavity and (b) an SEM (scanning electron microscopy) image of a single microplasma device with $50 \times 50\,\mu m^2$ emitting aperture (Becker et al., 2006; reproduced with permission).

(Paschen, 1889). Breakdown occurs when ions in the gas are accelerated across the gap so that a Townsend avalanche results (Townsend, 1925). For gaps greater than ~10 μm, breakdown has been well studied and occurs when the electric field becomes higher than ~3 V μm^{-1}. However, for air at atmospheric pressure an increase in breakdown voltage was observed for a gap spacing below ~5 μm (Schaffert, 1975). Under these conditions, electrons can tunnel through the surface potential barrier, a phenomenon called *field emission*. The field emission current is described by the Fowler–Nordheim equation (Fowler and Nordheim, 1928). Figure 5 shows the modified Paschen curve for air at 1 atm (Zhang et al., 2004a), in which the minimum and rise in breakdown voltage at small gap spacing is replaced with a plateau (the unmodified Paschen curve would have a minimum at ~5 μm gap spacing) and steep decline to zero. At the "knee" where the steep decline starts, field emission takes over completely from Townsend avalanche. The exact location of plateau and knee depends on the geometry, roughness, and composition of the metal electrodes. Finally, for ultrathin gaps smaller than ~2 nm, there is a finite probability that electrons can tunnel across an insulating barrier and result in a significant tunneling current.

Figure 5 indicates that breakdown in atmospheric pressure in very narrow gaps can occur at relatively low voltages, of a few tens of volts, which makes practical applications attractive. The situation becomes more favorable if one of the electrodes has a needle shape: The electrical field at the tip of the needle is intensified, and this leads to field emission at lower voltages than observed for the two planar electrodes of Figure 5. Work in our lab resulted in a factor 10 higher current density for a forest of

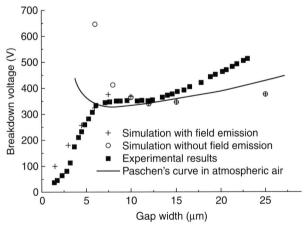

Figure 5 Paschen curve in atmospheric air, comparing simulation results with and without field emission, with experimental results (Zhang et al., 2004a; reproduced with permission).

tungsten oxide nanowires compared to a planar electrode at the same voltage (Ağiral et al., 2008a). Below we will show an example of the use of these nanowire electrodes in chemical synthesis. Another application of a combination of nanowires (carbon nanotubes (CNTs)) and a planar electrode is the recent work on miniaturized gas ionization sensors (Modi et al., 2003).

Figure 6 shows a typical field emitter design that consists of a sharp tip electrode and a *gate electrode*. This design is of the *Spindt type*, which refers to the fabrication process developed by Spindt et al. (1976). This design and others quite similar to it have become known for their application in plasma display devices, where the gate electrode acts as a counter electrode for the emitter electrode, and electrons emitted from the tip travel through a vacuum to hit a phosphorescent screen. The emission current at a given gate voltage strongly depends on the radius of curvature of the tip and the spacing between tip and gate, where the latter can be scaled quite easily by changing the gate hole opening. Reducing the size of this opening will give a higher emission current with a lower turn-on voltage. An example of this is a high-density array of field emitter tips (2.5×10^9 tips per cm^2) with an opening of 100 nm that can be driven by voltages of 10–20 V (Choi et al., 2001).

Field emitter tips with gate electrodes have not been applied for the generation of reactive species for chemical synthesis, and this may be an interesting field for future study, because of the extra flexibility that these devices may offer. Devices that go in this direction are the electron impact ionization sources used for mass spectrometers (Kornienko et al., 2000). A particularly interesting feature of gate emitters is the possibility of separating the region where electrons are generated (between the tip and the gate electrode) and the region where chemistry can be stimulated (just like in the plasma displays). One may think of focusing the electron beam on a catalyst surface to enhance surface reactions, or may use the space between tip and gate, if connected to an inlet hole which runs to the back of the glass base in Figure 6, to introduce a gas that will be

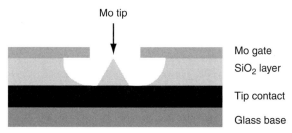

Figure 6   Field emitter tip of the Spindt type with an integrated gate electrode (Reuss et al., 2003; reproduced with permission).

ionized in or close to the gate opening. An exciting option would be a configuration in which each emitter-gate combination introduces a different gas. The high density of tips achievable may render the desired throughput of the system. Note also that if an array of individually addressable emitter tips is used, each emitter tip may be tuned to a different energy, allowing both a chemical gas-phase reaction to occur and ionization, if required at different locations and with a certain time interval.

2.2 Applications of microplasma reaction technology

Novel applications have been developed from the combination of microreactor technology and nonequilibrium microplasma chemistry. Here we discuss a selection from the recent literature on this topic to illustrate several main trends. We will focus on microplasmas in confined microchannels for the purpose of chemical synthesis and environmental applications.

2.2.1 Nanostructure synthesis in microplasma reactors

Synthesis of nanostructures using microplasma reactors is an attractive method since decomposition of the source material and subsequent crystal nucleation can be performed in the high-density nonequilibrium plasma within time intervals on the order of milliseconds. For example, Sankaran et al. (2005) synthesized silicon nanoparticles, 1–3 nm in diameter, from a mixture of argon/silane in a continuous flow atmospheric pressure microplasma reactor. Their technique is based on high pressure microdischarges with very short operation time (μs–ms). Microdischarges were created in a hollow cathode, which consists of a stainless-steel capillary tube with 180 μm ID (inner diameter) and extended toward an anode, a metal tube with 1 mm ID, as shown in Figure 7. Using a direct current microplasma which was sustained at 300–500 V and 3–10 mA, silicon nanoparticles were produced as an aerosol around atmospheric pressure. Since this microreactor operates at low powers (5–10 W) in plasma volumes less than 1 μl, resulting power densities were as high as $10 \, kW \, cm^{-3}$. Such a high-density plasma allows fast plasma processing for the synthesis of blue luminescent silicon nanoparticles. The high density of energetic electrons in the microdischarges efficiently decomposed the gaseous precursor to produce radicals in the reaction zone. At a radical concentration high enough for nucleation, nanoparticles can start to grow in the microplasma. When the particles are removed by the gas flow from the discharge zone to a zone with a low concentration of radicals, particle growth will stop. An additional feature of the system is that particle charging in the microplasma reduces coagulation downstream of the reaction zone. Using the

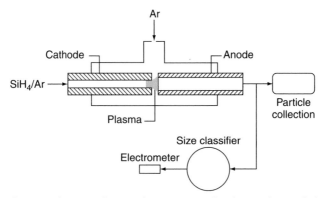

Figure 7 Schematic diagram of microplasma reactor for the synthesis of silicon nanoparticles. A microdischarge forms at the cathode tip and extends a short distance toward the anode (Sankaran et al., 2005; reproduced with permission).

same direct current microplasma technique, Fe and Ni catalyst particles were synthesized in a controlled way at atmospheric pressure and used for gas-phase growth of CNTs (Chiang and Sankaran, 2007, 2008). The catalyst particles were prepared from ferrocene and nickelocene. In summary, this simple and inexpensive microreaction technique can be used to synthesize nanoparticles in a continuous flow from the decomposition of gaseous precursors.

Nozaki et al. (2007a, 2007b) developed an atmospheric pressure microplasma reactor for the fabrication of tunable photoluminescent silicon nanocrystals (3–15 nm). They generated a capacitively coupled nonequilibrium plasma in a capillary glass tube with a volume of less than 1 µl and a residence time around 100 µs and used it to decompose silicon tetrachloride into atomic silicon. In the reactor a mixture of argon, hydrogen, and silicon tetrachloride was activated using a very-high-frequency (VHF, 144 MHz) power source. A schematic diagram of the experimental setup and an image of the microplasma reactor are shown in Figure 8. The upper electrode is connected to the VHF source (35 W discharge power) through a matching circuit and metallic electrodes with a 2 mm gap between them are around the outside of the capillary tube (borosilicate glass: ID = 630 µm, OD = 1,100 µm). Optical emission spectroscopic characterization of the microplasma indicated an electron density of $10^{15}\,cm^{-3}$, an argon excitation temperature of 5,000 K, and a rotational temperature of 1,500 K. Under these high-density reactive conditions, efficient decomposition of the silicon source gas and formation of a supersaturated silicon vapor lead to nucleation of gas-phase crystals via three-body collisions and subsequent rapid termination of crystal growth due to the very short residence time in the microreactor.

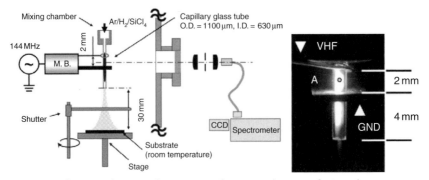

Figure 8 Schematic diagram of experimental setup and image of microplasma reactor with VHF source developed for the synthesis of photoluminescent silicon nanocrystals at room temperature (Nozaki et al., 2007a; reproduced with permission). M.B. is a matching electrical circuit.

An inductively coupled microplasma reactor was developed by applying ultrahigh frequency (UHF) to deposit a material on different substrates (Bose et al., 2006; Shimizu et al., 2003, 2005). An atmospheric pressure O_2–Ar microplasma reactor was used to prepare molybdenum oxide nanoparticles using molybdenum wire as the source material. The molybdenum metal wire with a diameter of 100 μm was inserted 6 mm from the exit of a pinched nozzle with an exit opening with an ID of 60–70 μm. A 20-turn copper coil was used to connect the reactor to the UHF source via a matching circuit. A drawing of the capillary microreactor is shown in Figure 9. A high-density

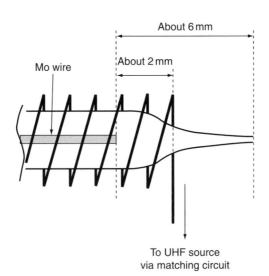

Figure 9 Capillary head of UHF microplasma reactor developed for the synthesis of molybdenum oxide nanoparticles (Bose et al., 2006; reproduced with permission).

microplasma oxidized the molybdenum and MoO_x nanoparticle seeds were supplied from the wire. The flow rate of the O_2–Ar mixture affected the process mechanism and therewith controlled shape, size, and oxidation state of the nanoparticles. It was found that MoO_2 oxidized to MoO_3 and that nanoparticle size decreased with an increase in flow rate. Using the same UHF microplasma technique, a tungsten wire was coated with multi-walled carbon nanotubes (MWCNTs) by flowing methane and vaporized ferrocene gases through a microchannel (Shimizu et al., 2005). This new microreaction method is unique as it gives a higher growth rate of MWCNTs with low power consumption (a few tens of watts) than conventional plasma-enhanced CVD processes.

Atmospheric pressure microplasma technology has the advantage of creating high-density reactive media in small spaces which can be used for surface modification and material deposition inside microchannels. We have recently developed a DBD technique to activate a coating of nickel/alumina catalyst in a capillary microreactor to enhance carbon nanofiber (CNF) growth on this coating (Ağiral et al., 2009). CNFs are promising nanostructured catalytic supports for liquid-phase reactions due to their high porosity and tortuosity (De Jong and Geus, 2000). Although thermal activation is an important way to significantly increase nanofiber yield, an atmospheric pressure microplasma may form an alternative route by using discharge-activated species which react at the catalyst surface at ambient temperatures. In our work, the fused silica capillary microreactor (500 µm ID, 550 µm OD) with an internal nickel/alumina catalyst coating was connected to a gas supply through a graphite ferrule high-voltage electrode. A DBD was generated to activate the catalyst at 300 K under a flow of hydrogen (5 ml min^{-1}) and helium (150 ml min^{-1}) for 15 min. The microreactor, catalyst coating, and microplasma treatment are shown in Figure 10. The catalyst color changed from light gray to dark gray after activation for 15 min showing that reduction of nickel took place during the discharge operation. Optical emission spectroscopic characterization showed that low-temperature activation of the catalyst occurs via active plasma species in the microreactor at atmospheric pressure. The discharge generated in the microchannel was characterized as uniform and stable with a high-power density (680 W cm^{-3}) at ambient gas temperature. The discharge treatment increased the CNF yield significantly compared to a nonactivated sample and the process can compete with a high-temperature treatment at 973 K for 2 h. Additionally, a comparison of the low-temperature microplasma treatment with a thermal treatment showed that the diameter of nanofibers is much more uniform in the former case. The method demonstrates the feasibility of cold catalyst activation on microreactor walls.

Another example of inner wall modification of microchannels with a microplasma is the deposition of uniform platinum films in microchannels

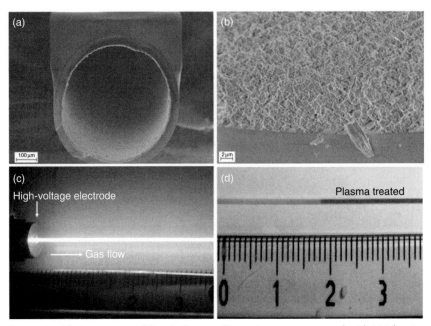

Figure 10   (a) SEM images of fused silica capillary microreactor coated with Ni/alumina catalyst, (b) catalyst layer after calcinations, (c) photo taken during microplasma treatment to increase catalyst activity for CNF synthesis, and (d) change in color of catalyst from light gray to dark gray after activation for 15 min (Ağıral et al., 2009; reproduced with permission).

(Kadowaki et al., 2006). This was done by generating a DBD at a low-pressure (a few Torr) in a capillary and in a microchannel in a glass chip, with electrodes attached to the outer surface along the channel axis. Photographs of the microplasma in the capillary and in the Pyrex chip are shown in Figure 11. Cylindrical graphite and metal foil electrodes were used for the capillary and the Pyrex chip, respectively. By introducing vaporized platinum bisacetylacetonate, plasma deposition led to a platinum film with a thickness of more than 100 nm. By controlling the voltage and frequency parameters, it was possible to achieve uniform deposition between the electrodes in the microchannel. This technique is a good example of the possibility of using microplasma technology to deposit a thin-film catalyst or other coatings in a controlled way in microreactors.

2.2.2 Environmental applications of microplasma reactors
Through the generation of highly reactive species such as energetic electrons and active radicals, microplasma reactors create novel process windows for C–C and C–H bond cleavage involved in the decomposition of harmful gaseous pollutants at atmospheric pressure. As an example of

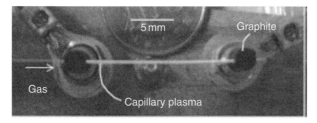

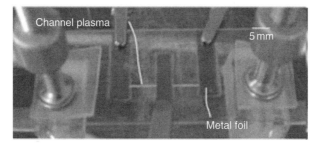

(a) Ar, 2 Torr

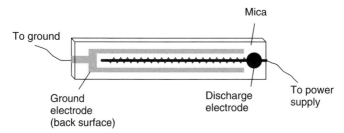

(b) He, 5 Torr

Figure 11 Photos of a microplasma in (a) a capillary and (b) a microchannel in a Pyrex chip, developed for plasma CVD of platinum films (Kadowaki et al., 2006; reproduced with permission).

this, a miniature DBD device was developed for the decomposition of volatile organic compounds (VOCs) (Seto et al., 2005). The device works with a surface discharge microplasma and contains microelectrodes manufactured by photoetching of stainless-steel foil and a dielectric substrate, acting as a barrier, made from a rectangular sheet of mica (Figure 12). By applying a high-voltage (3.5 kV) alternating current (AC) field (67 kHz) to the discharge electrode, a microplasma was formed on the surface of the mica sheet and high-energy electrons were generated which dissociated molecules, formed negative and positive ions, and excited molecular and

Figure 12 Illustration of a surface-discharge microplasma reactor developed for the decomposition of VOCs in the gas phase (Seto et al., 2005; reproduced with permission).

elemental species. Ion counting measurements showed that most of the by-products were negatively charged. The efficiency of toluene decomposition was found to be more than 99% in batch and 30–80% in continuous flow, and it was shown that toluene was completely converted into carbon dioxide by the atomic oxygen generated in the microplasma reactor.

Decomposition of tetrafluoromethane at atmospheric pressure was achieved with a microreactor which has very small electrode gaps (70 μm) between microstructured electrodes with an interdigitated arrangement (Figure 13) (Sichler et al., 2004). The merits of this reactor are low ignition voltages and a homogeneous plasma at high pressure. Alumina substrates, nickel electrodes, and Foturan® glass with an alumina coating were used as the microreactor materials. It was shown that micromachined flow structures provide effective flow control and have a large effect on decomposition efficiency. Additionally, scale-up to larger exhaust gas flows was achieved by "numbering up," that is, by constructing a multireactor with 16 microplasma reactors in parallel. Besides the larger throughput, the transition to a multireactor concept reduced the power strain on single microreactors and prolonged their lifetime. It was suggested that 25 of such multireactors are needed to treat 20 liter per minute of fluorinated waste gas for a small semiconductor plant, at an energy consumption of only 50% of that of a conventional combustion system.

Mori et al. used capillary discharge tubes with an ID of 0.5 or 3 mm to decompose carbon dioxide. The setup is shown in Figure 14 (Mori et al.,

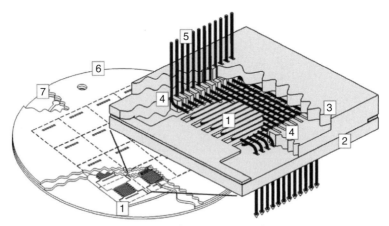

Figure 13 Illustration of a single microplasma reactor and its integration in a multireactor. Numbered features are (1) plasma source, (2) glass structure, (3) reaction chamber, (4) inlet and outlet of the reactor, (5) gas flow, (6) 4 × 4 array in a multireactor, and (7) contact pads for RF power (Sichler et al., 2004; reproduced with permission).

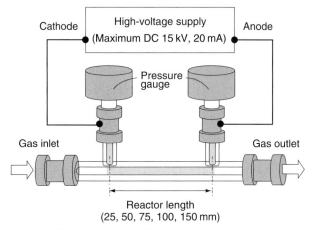

Figure 14   Schematic diagram of the capillary plasma reactor developed for the decomposition of carbon dioxide (Mori et al., 2006; reproduced with permission).

2006). The capillary plasma reactor consists of a Pyrex glass body and mounted electrodes which are not in direct contact with the gas flow in order to eliminate the influence of the cathode and anode region on CO_2 decomposition. Analysis of downscaling effects on the plasma chemistry and discharge characteristics showed that the carbon dioxide conversion rate is mainly determined by electron impact dissociation and gas-phase reverse reactions in the capillary microreactor. The extremely high CO_2 conversion rate was attributed to an increased current density rather than to surface reactions or an increased electric field.

The application of nanostructures as electrodes in a microplasma reactor was used to increase the reactivity and efficiency of barrier discharge processes at atmospheric pressure (Ağiral et al., 2008b). CNFs and tungsten oxide ($W_{18}O_{49}$) nanowires were integrated into a continuous flow microplasma reactor so that charge injection from the nanostructures by field emission supplied free electrons and ions after discharge. Incorporation of the nanostructures was performed by growing nanowires and nanofibers on the silicon chip which was used as a high-voltage electrode in a glass microreactor system, as shown in Figure 15. Atmospheric pressure field electron emission tests showed that field enhancement at the tip apex of the nanostructures results in electron emission in air. Injection of charged species during discharge generation results in a decrease in breakdown voltage and a higher power deposition, at the same measured potentials as applied on electrodes without nanostructures. As a model reaction, CO_2 cracking was tested and it was found that the chemical reactivity of the discharge is increased by application of the nanofibers.

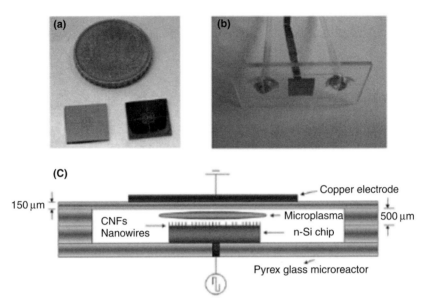

Figure 15 On-chip microplasma reactor using nanostructured electrodes, (a) silicon chip before and after a CVD process for nanostructure growth, (b) microplasma reactor, and (c) general diagram of the device (Ağiral et al., 2008b; reproduced with permission).

2.2.3 Chemical synthesis in microplasma reactors

Performing plasma processes in a continuous-flow microreactor leads to precise control of residence time and to extreme quenching conditions, therewith enabling control over the composition of the reaction mixture and product selectivity. In a nonequilibrium microplasma reactor, low-temperature activation of hydrocarbons and fuels, which is difficult to obtain in conventional thermochemical processes, can be achieved at ambient conditions.

Nozaki et al. (2004) described the application of a microplasma reactor in partial oxidation of methane. The plasma generation principle in this case is similar to a DBD and gives high-energy electrons which activate methane–oxygen mixtures for direct production of methanol. The micro-reactor consists of a Pyrex thin glass tube (ID $= 1.0$ mm, length $= 60$ mm) with a twisted metallic wire (ID $= 0.2$ mm, length $= 100$ mm) inside, as shown in Figure 16. Power consumption was calculated to be between 3 and 10 W. Excess heat generated by partial oxidation was efficiently removed from the microreactor, and successive destruction of formed oxygenates was minimized in the highly quenching environment. It was possible to produce methanol reproducibly in a one-pass process, with 10% maximum yield at room temperature, and at 100 kPa within 280 ms without explosion of the methane/oxygen mixture. The advantage of

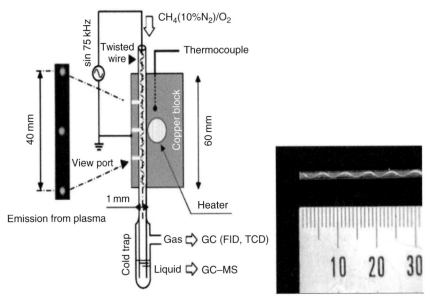

Figure 16 Microplasma reactor setup for partial oxidation of methane (left) and photo of thin glass tube equipped with a twisted metal wire (right) (Nozaki et al., 2007a; reproduced with permission).

using a plasma microreactor for fuel processing is that methane can be activated by high-energy electrons to achieve 40% conversion, independent of temperature and pressure. Additionally, unlike in thermochemical reactions, product selectivity is independent of methane conversion. However, the present microplasma reactor has not yet been made compatible with existing methanol manufacturing processes, and the reactor dimensions and power consumption need to be optimized in terms of a balance between excitation and quenching processes.

Hsu and Graves (2005) have used a micro hollow cathode as a microreactor to decompose ammonia and carbon dioxide. An MHCD can provide a highly reactive environment with a high electron temperature, power density, and ion density which would be ideal for endothermic cracking reactions. Decomposition of ammonia into nitrogen and hydrogen can be used as a source of pure hydrogen, while cracking of carbon dioxide can be used to dispose of radioactive carbon dioxide, or for the production of oxygen. In this case, the MHCD was constructed from two molybdenum electrodes (100 μm thick) sandwiching a mica dielectric (260 μm thick). The three layers were glued together and a 200 μm hole was drilled to construct a continuous flow microreactor. Significant decomposition of ammonia and carbon dioxide with effective reaction temperatures exceeding 2000 K was shown. As a demonstration of the numbering-up principle, it was shown that with two or more microreactors in series the conversion could be increased significantly. This work demonstrated that microplasma-induced generation of hydrogen from ammonia in a flow-through MHCD is feasible, however, to become of economic relevance, the overall power efficiency should be increased. This may be done by pulsing the plasma, or operating many microplasma reactors in parallel and/or in series. It was proposed to pulse the discharge with short microsecond pulses to minimize electrical power input and stabilize the plasma.

Direct hydroxylation of benzene to phenol and of toluene to cresol in a microplasma reactor was carried out using a DBD at atmospheric pressure (Sekiguchi et al., 2005). This type of discharge provides hot electrons which dissociate molecules and therewith initiate hydroxylation reactions at ambient gas temperature. The glass microreactor studied by Sekiguchi et al. has a rectangular shape (100 mm in length and 70 mm in width) with aluminum electrodes and Teflon sheets as spacers, see Figure 17. Energetic electrons and oxygen radicals can dissociate the aromatic ring and functional groups, which is followed by oxidation. It was proposed that the selectivity and the yield of this plasma-based direct hydroxylation process

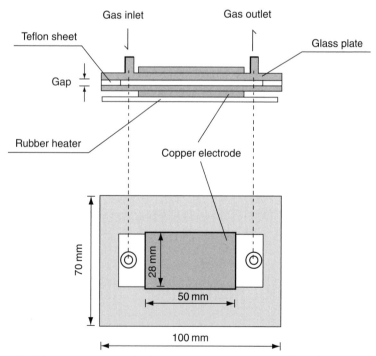

Figure 17   Schematic diagram of microplasma reactor, based on DBD, for hydroxylation of benzene and toluene (Sekiguchi et al., 2005; reproduced with permission).

may be increased by controlling the production of oxygen radicals via changes in the reactor surface or plasma generation methods.

Oxidative conversion of light alkanes, C_1–C_3 range, was carried out in a dielectric barrier-type microplasma reactor (Trionfetti et al., 2008a). The direct conversion of alkanes is largely obstructed by the strong C–H (415 kJ mol^{-1}, for methane) and C–C bonds (350 kJ mol^{-1} for ethane) (Choudhary et al., 2003). Cold plasma processing can be an alternative to high-temperature thermochemical processes. A barrier discharge treatment in a confined reactor offers the advantages of a uniform and dense plasma with a better control of residence time. The tested microplasma reactor (30 mm length, 5 mm width, and 500 μm channel depth) was fabricated by thermal bonding of three Pyrex layers and attaching copper electrodes on top and bottom at the outside of the chip, as shown in Figure 18. A plasma was generated by applying a high-voltage (5–10 kV) sine wave (60 kHz) to the top electrode while the bottom electrode was grounded. Heat produced during the oxidative conversion of alkanes was easily removed due to the small volume and high surface area of the microreactor, so that it operated at ambient temperature. The feed composition was 10% alkane and 1% oxygen in helium. Activation of hydrocarbons follows two main routes. In the first one, energized electrons dissociate alkane molecules by cleaving C–H and C–C bonds. The direct observation of CH, C_2, and H excited species by an optical emission spectrometer is an indication of this bond cleavage at room temperature. Secondly, electron impact dissociation of oxygen molecules produces active oxygen radicals which initiate radical chain reactions. Excited helium species also may play a role in the process, by transferring energy to alkane and oxygen molecules. In the experiments with propane, a high selectivity (37%) to products with a molecular weight higher than propane (C_4, C_4^+) was observed, indicating that under microplasma conditions C–C bond formation occurs. Coupling reactions between radicals are favored at lower temperatures and the cold plasma process in a

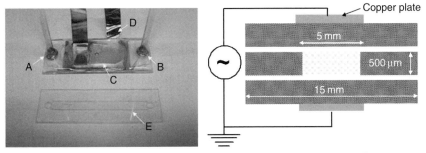

Figure 18   Microplasma reactor, based on DBD, for oxidative conversion of C_1–C_3 alkanes; A: gas inlet; B: gas outlet; C: frontside copper electrode; D: backside electrode (Trionfetti et al., 2008a; reproduced with permission).

microreactor could be an alternative for upgrading light hydrocarbons by direct homologation of alkanes. A kinetic model of plasma propane conversion in this microplasma reactor was developed for better understanding of physical and chemical processes over a range of reactor operation conditions (Ağiral et al., 2008c). The model employed the well-mixed plasma reactor module of Chemkin 4.1 (CHEMKIN, 2004) to determine the time-averaged species composition and electron energy balance which equates the rate of change of the electron swarm internal energy to the net flow of electron enthalpy into and out of the reactor, therewith accounting for net chemical production rates, surface losses, collisional losses, and power deposition from the externally applied electromagnetic field. Reaction rate coefficients of electron impact reactions with propane were determined with the aid of BOLSIG+ software (Hagelaar and Pitchford, 2005). This software also calculates the relation between the average electron energy and the reduced electric field inside the microreactor. The estimated average electron temperature was used to obtain a steady-state solution in the Chemkin plasma reactor model. The BOLSIG+ code uses the two-term spherical harmonic expansion of the electron energy distribution function to solve a zero-dimensional Boltzmann equation. Cross sections of partial dissociative excitation and ionization processes ($e^- + C_xH_y$, $x = 1\text{--}3$; $y = 1\text{--}8$) were obtained from the experimental data of total dissociation cross sections and of total cross sections for dissociative ionization (Janev and Reiter, 2004). It is necessary to correct the residence time since a barrier discharge consists of filaments, and this was done on the basis of a quantitative agreement with experimental data. The model includes electron impact dissociation and ionization, ion–neutral reactions, neutral–neutral chemistry, and surface recombination of ions at the walls. Simulated results were compared with experimental data and a good agreement was found. H, CH, CH_2, CH_3, C_2H_3, C_2H_5, C_3H_5, C_3H_7, and C_4H_9 radicals were found to play an important role during propane conversion in the microplasma reactor. At higher propane conversion levels, enhanced C–C bond formation was observed.

The same microplasma reactor was used to study the feasibility of oxidative dehydrogenation of propane in the presence of a Li/MgO catalyst (Trionfetti et al., 2008b). It was anticipated that a synergistic effect between catalytic and plasma processes may be obtained, that possibly may give a higher conversion and yield of target products. The reason for this expectation was based on the following: First of all, a catalyst supported on an insulator oxide deposited in the barrier discharge region may influence the plasma properties due to a change in surface properties and permittivity of the dielectric material, and this influence may be positive. A well-chosen catalyst will, as always, decrease the overall activation energy, but the selectivity of a catalytic reaction may be increased by selective plasma activation of specific molecular bonds between adsorbed species and surface. This can be thought to occur as follows: In a nonequilibrium plasma, molecules are excited by electron impact. Among the characteristic species (electrons, ions, molecular fragments and excites species such as electronic, vibrational, rotational/translational excitations), only radicals and vibrationally excited species will be relevant for surface reactions in an atmospheric pressure nonequilibrium plasmas, due to the fact that radicals have long relaxation times compared to the time needed for chemical reactions, plus that they have a high sticking probability on surfaces. Vibrational activation in the dissociation degree of freedom can lower the activation barrier for dissociative adsorption (Tas, 1995).

In addition to the above, in a microplasma reactor a more intensive interaction of plasma and catalyst surface can be achieved in a confined environment. Li/MgO catalysts in the presence of oxygen at high temperatures have $[Li^+O^-]$ defect sites which activate C–H bonds in alkanes (Wang and Lunsford, 1986). Propane activation starts with hydrogen abstraction by oxygen ions, forming propyl radicals (Balint and Aika, 1997), C–C and C–H bond cleavage happens at high temperatures $(T > 823\,K)$ in the presence of $[Li^+O^-]$ centers. At these temperatures, a loss of catalyst area was observed, which results in less heterogeneous formation of propene (Trionfetti et al., 2006). A microplasma reactor may allow initial propane activation at lower temperatures by enhanced radical surface interactions in the confinement of a microreactor. To test this, a Li/MgO catalyst was deposited on the surface of a glass microchannel by micropipetting a sol–gel precursor system. The catalytic microplasma reactor showed enhanced olefin selectivity in the presence of Li/MgO, which indicates the formation of defect sites at ambient temperatures. Formation of higher hydrocarbon products $(C_4 + C_4^+)$ showed that coupling of radicals occurs predominantly in the homogeneous phase.

Anderson et al. (2008) have fabricated plastic microreactors based on parallel linear arrays of interconnected cylindrical microcavity plasma devices, using replica molding in UV-curable polymers. Their study was aimed at on-chip plasma processing with the generation of gas or solid phase from a gas feedstock. Figure 19 shows a magnified view of the 10×10 arrays of $400\,\mu m$ diameter microplasma devices, operating in 600 Torr of Ar. Deposition of $(C–S)_n$ microstructured polymer was done in Ar/CS_2 plasma. This study showed the feasibility of using low cost and disposable polymer microplasma reactors for potential chemical synthesis applications.

2.2.4 Generation of plasma in a liquid or at a liquid interface

Although, as was discussed above, plasma and other discharges are usually discussed in the context of gas-phase processes, electrical discharges are also quite well studied in *liquids*. For example, liquid-phase discharge reactors have recently been applied in drinking water and

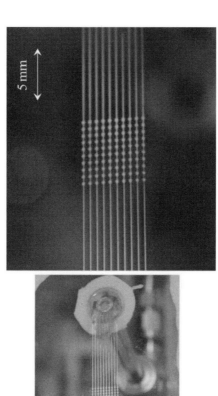

Figure 19 Photographs of microplasma reactor fabricated by replica molding on a plastic substrate (left) and magnified view of the 10 × 10 array of 400 μm diameter microcavity plasma channels, operating in argon (right) (Anderson et al., 2008; reproduced with permission).

wastewater treatment (Locke et al., 2006). Electric fields applied to water, called *electrohydraulic discharge*, initiate both chemical and physical processes, with two basic discharge types, differing by the amount of energy deposited in the system. The *corona* system uses discharges of ~1 J per pulse and operates at high frequency, 100–1000 Hz, with a peak current below 100 A and nanosecond voltage rise times. The *pulsed arc* discharge uses energy of ~1 kJ per pulse and larger, and operates at low frequency, 10^{-2}–10^{-3} Hz, with a peak current above 1 kA and microsecond voltage rise. The pulsed arc generates strong shock waves accompanied with cavitation and bubbles with ionized gas inside, leading to chemical reactions that have been suggested to be similar to those that occur in sonochemistry. In the corona discharge a streamer is generated, relatively weak shock waves are formed and a moderate amount of bubbles is observed. Radicals and reactive species are formed in the narrow region near the corona discharge electrodes: Generation of H, O, and OH radicals were observed by emission optical spectroscopy (Locke et al., 2006).

Corona discharges in liquids can be generated in many different reactor configurations (Locke et al., 2006), but principally they fall into two categories (Figure 20), one in which both electrodes are immersed in the liquid, and one in which one of the electrodes is above the liquid. For the second configuration, ions, radicals, and neutral species produced by the discharge, which partially occurs in the gas phase, may transfer into the liquid phase through action of the electric field, and react there to form other reactive species. It was suggested that the average energy of the positive gaseous ions entering the liquid phase may be more than 100 eV.

Note that the configurations in Figure 20 involve *point electrodes*. To initiate a pulsed discharge in water, it is necessary to have a high-intensity electric field (10^7–10^9 V m^{-1}) at the tip of the electrode. Proper insulation of the electrode is essential, because water is much more conductive than air. A small protrusion of the point electrode from the insulator surface

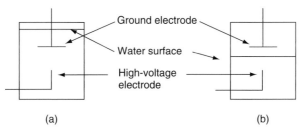

(a)　　　　　　　　　　　　　　　(b)

Figure 20　Schematic of two configurations for a pulsed corona reactor: (a) ground electrode submerged in liquid and (b) ground electrode suspended above the liquid surface. The immersed electrode in both cases is a point electrode (reprinted with permission from Grymonpré et al., 2004; Copyright 2004 American Chemical Society). For a variety of other reactor configurations, see Locke et al. (2006).

(less than 1 mm) gives a better discharge, because the electric field concentrates near the electrode edge or interface between the metal and the insulator. In air or gas insulation is less important, because the air surrounding the needle is a very good insulator.

As an example of an application of a gas–liquid pulsed corona discharge reactor, Grymonpré et al. utilized high-voltage needlepoint electrodes submerged in water coupled with a planar ground electrode in the gas phase above the water to remove low concentrations of phenol (Grymonpré et al., 2004). They found that the liquid-phase discharge leads to the formation of hydrogen peroxide and hydroxyl radicals, and the gas-phase discharge leads to the formation of ozone. A reticulated carbon electrode produced a higher number and more uniform distribution of streamers in the gas phase, which gave a higher amount of ozone dissolved in the liquid phase.

Other examples of gas–liquid electrical discharges are the degradation of 4-chlorophenol in water with bubbling air, in a multipoint-plate pulsed high-voltage reactor (Lei et al., 2007), and flue gas desulfurization by corona discharge in a cylindrical reactor with a wetted lining (Jiang et al., 2006) (*note that this is a typical microreactor configuration*). Corona discharge with a wet interface, but in a different reactor design, has been developed for industrial-scale desulfurization (Yan et al., 2006). The concept is based on a partitioned wet reactor system in which SO_2 in the flue gas is absorbed with ammonia water in a repeated spraying process, and then transferred at an appropriate flow rate to the successive plasma reaction stage in which uniformly distributed streamers are generated, in which the sulfites in the solution undergo plasma oxidization. The resulting output liquid is dried to produce a powder product using heat produced in the reactor. An interesting claim by the authors (unfortunately only mentioned in the abstract of the paper and not elaborated further) is that SO_2 absorption by the liquid is enhanced by *electric wind*, a concept to be discussed in a following section.

Just as was the case for gas-phase plasma devices, miniaturization of gas–liquid discharge reactors has several advantages, the most important being the possibility to keep the voltages for initiation of streamers or other discharge characteristics low, and the possibility to work at high pressures (atmospheric or higher) instead of vacuum. As was discussed above, the best results for larger systems have been obtained for needle electrodes, which one may also consider a step toward miniaturization.

One particular example of what may be called a gas–liquid discharge microreactor is the work of Baba et al. (2006, 2007), who have demonstrated the generation of an atmospheric pressure glow-discharge plasma in contact with liquid paraffin, using a capacitively coupled plasma method. The choice for paraffin has two reasons: no hydroxyl group present (which is thought to capture electrons, as is the case in water) and low volatility. Two types of plasma source, a mesh electrode-type and a parallel wire electrode-type, were applied, with typical electrode distance of 500 μm for the mesh and 1–5 mm for the parallel wire. The configuration was of the type (b) in Figure 20, with the mesh or parallel wires immersed in the liquid. Optical emission measurements showed evidence of CH and C_2, originating from paraffin, while laser-Raman scattering spectroscopy revealed that a graphitic soot-like material is produced.

Another microfluidic example in which a plasma is generated at a liquid–gas interface, is the work on a miniaturized discharge device for atomic emission, called electrolyte as a cathode discharge . This concept was developed to solve the problem of achieving adequate sample transport from the liquid to the gas phase, for analysis in a plasma detector comparable to the earlier analytical gas-phase microplasma devices (Eijkel et al., 1999). Conventional emission spectroscopy techniques achieve liquid sample introduction by nebulization of some sort or by drying the sample on one of the discharge electrodes, but such methods are not easily applicable in miniaturized analytical plasma devices. The desire remains to analyze a continuous liquid stream from a microchannel, as it would allow low-dead-volume in-line analysis of fractions from, for example, a microfluidic separation column. Figure 21 shows a schematic drawing of the developed device (Jenkins et al., 2005). Although this was meant to be used only for analysis, one can easily imagine using such a device (in a parallel format, to generate throughput) for the purpose of chemical synthesis.

Probably because the field is relatively young, not much has been reported yet on gas–liquid discharges in a microfluidic (microreactor) format. Yamatake et al. have used a DC-driven atmospheric plasma micro hollow cathode to directly inject O radicals into a fast oxygen gas

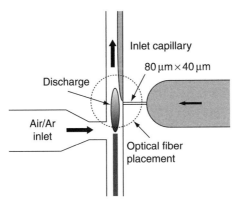

Figure 21 Schematic of aqueous electrode chip design (Jenkins et al., 2005; reproduced by permission of the Royal Society of Chemistry).

stream which was introduced in a solution of acetic acid, as a model for water treatment. A clear correlation was found between acetic acid decomposition rate and gas-flow velocity, which indicates that rapid injection of the key radical, which is thought to be O and which has a short lifetime in atmospheric oxygen, at the gas–liquid interface is crucial (Yamatake et al., 2006).

A interesting option would be the use of segmented flow (Taylor flow) patterns in a microchannel (Günther et al., 2004) in combination with a DBD as described above. Effectively, such a configuration may lead to similar chemistry as in a sonochemical process, but probably with a much higher efficiency, since the gas holdup volume in a microchannel with Taylor flow can be made much higher. An example of a discharge accomplished over a bubble train, but based on a conventional bubble column (see Figure 22) has recently been reported (Katayama et al., 2009).

A final topic to be discussed in this section is the direct injection of electrons into a liquid by the use of *nanowires*. The difference with the discharge processes described above for the point electrodes is twofold: the use of *nanostructured electrodes* and the generation of *solvated electrons*, instead of the initiation of a discharge. Solvated electrons have a very short

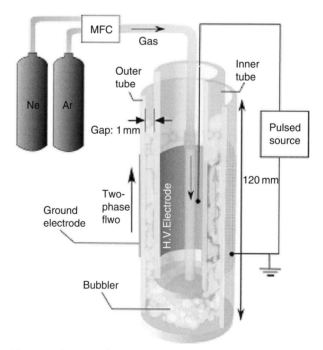

Figure 22   Schematic diagram of a DBD reactor, in which a discharge is carried out over a bubble column (from Katayama et al., 2009; © 2009 IEEE).

lifetime in water, on the order of 50 fs, due to the fast response of the hydrogen bond network (Bragg et al., 2004), but they may be stable for hours to months in other liquids, like liquid NH_3 (Jortner and Kestner, 1973).

In very pure nonpolar dielectric liquids, electron injection currents at very sharp tips follow the Fowler–Nordheim voltage dependence (Halpern and Gomer, 1969), just as is the case in solid insulators, and in a gas, as described before. In a study of the electrochemical behavior of CNT cathodes (Krivenko et al., 2007) direct experimental proof was found of electron emission into the liquid hexamethylphosphortriamide, which was chosen because it is a convenient solvent for the visualization of solvated electrons at room temperature: the solution will show an intense blue coloration upon the presence of solvated electrons. Electron spin resonance showed prove of a free electron. Electrogenerated (as opposed to photogenerated) solvated electrons have been used in the synthesis of L-histidinol (Beltrá et al., 2005), albeit that in that work the electrons were generated electrochemically from a solution of LiCl in $EtNH_2$, which is a solvent that is easier to handle than liquid ammonia (boiling points at atmospheric pressure are 17 °C and –33.34 °C, respectively).

These results suggest that the use of nanofiber electrodes in a microreactor environment to generate solvated electrons for chemical synthesis, may offer an interesting new route for reduction reactions. We are currently working on this concept in our laboratory (Ağiral et al., 2010).

3. ELECTROCHEMICAL MICROREACTORS

Electrochemistry, to distinguish it from the topics discussed in previous sections, is concerned with *low-energy charge transfer in solution*. The electron transfer typically occurs on the surface of a charged (usually metal) electrode. Possible chemical reactions that may occur, and that may be of importance in chemical synthesis, are the generation or annihilation of gases (in an electrolysis or a fuel cell, respectively) and the generation or neutralization of ions, which may be accompanied with the dissolution or deposition of a solid material.

The reasons to perform electrochemistry, in particular, *electrosynthesis*, in a *microfluidic system* are the following (Rode et al., 2009): (1) reduction of ohmic resistance in the electrochemical cell, by decreasing the distance between anode and cathode, (2) enhancement of mass transport by increase of electrode surface to cell volume ratio, also realized by small interelectrode gaps, (3) performing flow chemistry to establish single-pass conversion, and (4) coupling of cathode and anode processes, permitting simultaneous formation of products at both electrodes. The latter possibility is very interesting from a sustainability point of view (less waste) and examples have started to appear in recent literature, like the epoxidation of propylene (Belmont and Girault, 1995). Paddon et al. have recently reviewed paired and coupled electrode reactions for electroorganic synthesis in microreactors (Paddon et al., 2006).

A fifth reason for using microfluidics in electrochemistry would be the possibility to combine flow chemistry with an ultrafast mixer, which allows the generation and subsequent use of short-lived reactive ions or radicals, for example, in a "cation flow" process (Suga et al., 2001; Yoshida, 2008). Finally, a sixth reason for performing electrochemistry in a microfluidic system may be the desire to efficiently remove reaction heat (or joule heat due to high currents in combination with a high ohmic resistance) in fast electrochemical reactions (Yoshida, 2008).

The research on electrochemistry in microreactors has been reviewed in a number of recent publications (Hessel et al., 2004; Rode et al., 2009; Yoshida, 2008; Yoshida et al., 2008); therefore, we do not want to go into too much detail here. But since these reviews almost exclusively concern electroorganic synthesis, a number of other applications will be highlighted here.

A specific type of electrochemical microsystem, although not directly a synthetic microreactor, is a *microfluidic fuel cell*. In a recent review (Kjeang et al., 2009), the developments and challenges in this field have been analyzed. Kjeang et al. define a microfluidic fuel cell as a fuel cell with fluid delivery and removal, reaction sites, and electrode structures all confined to a microfluidic channel. Most microfluidic fuel cells use colaminar flow, without a physical separation (like a membrane) between anode and cathode. The laminar flow characteristic is utilized to delay mixing of fuel (the anolyte) and oxidant (the catholyte), which occurs by diffusion only and is restricted to a small zone close to the anolyte–catholyte interface at the center of the channel. Electrodes are positioned on the walls of the channel, at sufficient distance from the diffusion zone in order to prevent fuel crossover. Both anolyte and catholyte contain supporting electrolyte which facilitates ionic transport. Figure 23 shows a specific example of a microfluidic fuel cell, namely one in which an air breathing cathode is used to eliminate one of the main problems with microfluidic fuel cells, that is, the mass transport limitation at the cathode in cases where dissolved oxygen is used as the oxidant.

The colaminar configuration has a number of advantages over other fuel cell designs: (1) mixed media operation, (2) operation at room temperature, and (3) no auxiliary humidification, water management, or cooling systems. Additional advantages may arise from the specific possibilities that micromachining offers, for example, special electrode designs like tapered electrodes to accommodate the downstream growth of the mixing zone and maximize fuel utilization (Bazylak et al., 2005) or

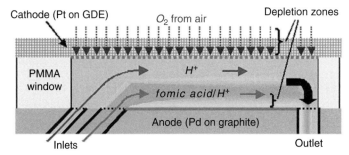

Figure 23 Air-breathing microfluidic fuel cells showing the colaminar flow principle, in combination with oxygen capture via gas diffusion through a porous cathode A three-phase interface is established between gas, electrolyte, and catalyst/solid electrode (reprinted with permission from Jayashree et al., 2005. Copyright 2005 American Chemical Society).

inlet and outlet manifolds for active control of concentration boundary layers by removing consumed species or adding reactants, respectively (Yoon et al., 2006). Despite all these advantages, current microfluidic fuel cells still have a too low-energy density (i.e., energy output per system volume or mass) to be of real practical value, which is mainly due to the single-pass use of electrolyte. An important issue in flow electrochemistry is the elimination of the supporting electrolyte that is added to enhance ionic transport. This electrolyte causes separation problems and product contamination. One option to minimize supporting electrolyte is to use a two-phase flow microfluidic system in which the reagent flow and the supporting electrolyte flow are kept separate. An example is the use of a two-phase flow with N-octyl-2-pyrrolidone (NOP) and aqueous electrolyte and with acetonitrile and aqueous electrolyte, in contact with a polycrystalline boron-doped diamond electrode (MacDonald et al., 2009). Two different configurations are possible: The first applies *pre-electrolysis*, in which the aqueous phase is oxidized to form ozone, bromine, peroxocarbonate, or the-like, before contact with the organic reactant phase, while the second has an aqueous and an organic phase, which contains the electrochemically active reagent, simultaneously in contact with the working electrode. Only a thin reaction zone close to the *triple-phase boundary zone* is active during the electrochemical reaction. The high electrical resistance within the organic phase restricts the reaction zone. Both configurations give clean electro-synthesis without electrolyte included in the organic phase, but the acetonitrile system gave better results due to its lower viscosity, which enhances mass transport and conversion rate, and its more stable triple-phase boundary electrolysis zone, with virtually no *undercutting* of the organic phase under the aqueous phase like it has been observed for NOP. Undercutting is a result of surface tension changes during electrochemistry (MacDonald et al., 2007). Electrochemistry at static triple-phase boundaries was also demonstrated using droplets (Banks et al., 2003; Marken et al., 1997; Scholz et al., 2005), porous host electrodes (to optimize triple-phase boundary zones) (Ghanem and Marken, 2005; Niedziolka et al., 2007), microwire-based electrodes (Bak et al., 2007), and microdroplet arrays on lithographically modified electrode surfaces (Rayner et al., 2007).

A similar distinction between a system with pre-electrolysis with only one electrode (in this case anodic) process, and a system with simultaneous anodic and cathodic processes (in which anode and cathode are on opposite walls of a microchannel so that each liquid is only in contact with the desired electrode potential, analogous to the fuel cell configurations discussed above) was made by Horii et al. (2008) in their work on the *in situ* generation of carbocations for nucleophilic reactions. The carbocation is formed at the anode, and the reaction with the nucleophile is either downstream (in the pre-electrolysis case) or after diffusion across the liquid–liquid interface (in the case with both electrodes present at opposite walls). The concept was used for the anodic substitution of cyclic carbamates with allyltrimethylsilane, with moderate to good conversion yields without the need for low-temperature conditions. The advantages of the approach as claimed by the authors are efficient nucleophilic reactions in a single-pass operation, selective oxidation of substrates without oxidation of nucleophile, stabilization of cationic intermediates at ambient temperatures, by the use of ionic liquids as reaction media, and effective trapping of unstable cationic intermediates with a nucleophile.

An interesting development is the coupling of an electrochemical microreactor with a continuous separation process. Performing a synthetic chemical process in a continuous flow microreactor is now well established, but in the end the benefits of continuous processing will be lost if not also the downstream work-up processes can be performed in a continuous fashion. That is why at the moment a large research effort is devoted to coupling (or integrating) separation methods with microreactors (see Hartman and Jensen, 2009 for a recent review). An example is the coupling of a microreactor for electroorganic synthesis, configured as an electrochemical thin layer cell, to a *simulated moving bed* (SMB) separator (Küpper et al., 2003; Michel et al., 2003). SMB technology is a preparative chromatographic method which is turned into a continuous process by simulating a counterflow of the adsorbing stationary phase (a particle bed) and the mobile phase, by leaving the adsorbing phase unmoved but switch the positions for inlet of eluent and feed, and for outlet of raffinate and extract (Ruthven and Ching, 1989). In the work of Küpper et al., the SMB process has not been implemented, but the electrochemical reactor is designed to be operating with a conventional SMB plant (Küpper et al., 2003). Michel et al. describe a combination of electrochemical reaction and chromatographic SMB separation applied to the direct electrochemical production of arabinose and simulate its operation. The electrochemical microreactors are placed between the chromatographic columns. The concept has the advantage of being able to switch the electrochemical reactors on and off by means of the applied current. Also in this work no experimental verification of the concept is given, but the models prove the feasibility of the integrated process, as higher yields than with a serial connection of reactor and SMB can be obtained, although the integrated process will obtain high productivity only at low yield and vice versa.

Micro-SMB separators have only been studied numerically (Subramani and Kurup, 2006), but one can think of ways to implement a *real* moving bed in a miniaturized version by applying a shifting magnetic field on, for example, magnetic resin beads or applying DEP on adsorbent particles, in a microchannel.

As a final topic for this section on electrochemical microreactors the electrochemical generation of cofactors in biocatalytic microreactors is worth mentioning. The use of immobilized enzymes has progressed rapidly in medical and analytical applications and in the food and beverages industry biocatalysts are used in a variety of processes and products. Immobilization of enzymes has the advantage over dissolved enzymes of easy recovery and reuse and (in many cases) improved stability. It is therefore not surprising that microreactors applying immobilized enzymes as a biocatalyst have received increasing attention in the recent literature.

For optimal operation enzymes may need the assistance of cofactors, for example, many of the dehydrogenases require nicotinamide adenine dinucleotide ($NAD(P)^+/NAD(P)H$). Because of the considerable cost of this cofactor, continuous regeneration is desired, which is possible by electrochemical regeneration from NAD^+ using a mediator (e.g., flavin adenine dinucleotide, FAD) which transports electrons from a cathode to an enzyme (e.g., formate dehydrogenase, FDH) which in its turn regenerates NADH from NAD^+. $FADH_2$ rather than formate serves as the substrate for the enzyme FDH, and enough $FADH_2$ can be generated at the electrode in order to shift the unfavorable equilibrium to NADH formation. However, due to the reverse reaction, which runs spontaneously at pH 7, the concentration of electrochemically generated $FADH_2$ remains low in the bulk solution in classical batch reactors, which is why *in situ* and local generation of $FADH_2$ is preferably performed in a microfluidic reactor (Yoon et al., 2005). The microreactor reported by Yoon et al. utilizes multistream laminar flow to focus a reagent-containing stream close to an electrode, by adjusting the flow rate ratio of reagent and buffer stream flowing in parallel. $FADH_2$ is indeed produced in sufficiently high concentrations at the electrode to drive the subsequent reaction toward NADH regeneration. Another example of the same mediator concept is the regeneration of NADH in a filter-press microreactor with electro-eroded cylindrical microchannels (Kane and Tzédakis, 2008). The high specific surface area of the cathode of $250 \, cm^{-1}$ in this microreactor provided the desired conditions to increase the $FADH_2/FAD$ concentration ratio and shift the nonspontaneous reaction $FADH_2/NAD^+$ toward regeneration of NADH. Both reactors (Kane and Tzédakis, 2008; Yoon et al., 2005) were tested for the synthesis of chiral L-lactate from pyruvate in the presence of L-lactate dehydrogenase.

A completely different example of cofactor generation in a microsystem is the local production of magnesium ions to control the position of activation of a DNA restriction enzyme (Katsura et al., 2004). Alternative DNA restriction schemes are important to cut DNA at specific positions, different from the conventional molecular biological methods. Control of restriction was achieved by applying a direct current to a needle electrode of magnesium. Only when and where the magnesium ions from this needle were produced, the restriction enzyme became activated.

4. ELECTROKINETIC CONTROL OF CHEMICAL REACTIONS

Instead of directly using the charged or otherwise electrically activated species in a chemical reaction, the option exists to use charges to *enhance mass transport*. This can be achieved by transporting the chemical species, if charged, themselves, or by transporting the medium in which the chemical species are contained. This *electrokinetic transport* exists in different forms, which will be highlighted below.

4.1 Electrophoresis and electroosmosis

In case an electric field is applied on an electrolyte in, for example, a glass tube, two different responses to the electric field will occur:

1. *Electrophoresis*: the positive ions in the electrolyte will move in the direction of the electric field E, while the negative ions will move in the opposite direction. The velocity v_{EF} by which the ions move is proportional to their electrophoretic mobility, μ_{EF}:

$$v_{EF} = \mu_{EF}E \qquad (1)$$

Electrophoretic mobilities are typically in the range of $3–8 \times 10^{-4}\,cm^2$ $V^{-1}s^{-1}$, exceptions being the H^+ and OH^- ions, with mobilities of 36.25×10^{-4} and $20.50 \times 10^{-4}\,cm^2\,V^{-1}\,s^{-1}$, respectively.

2. *Electroosmotic flow* (EOF), consisting of the motion of the liquid along the surfaces of the electrolyte container. This EOF is based on the drag force that is exerted on the electrolyte by double-layer charge moving in an electric field. The double layer is a result of a change in space charge density close to container walls which become charged when they are in contact with electrolyte (e.g., a glass wall will carry

a negative charge in contact with an aqueous liquid at neutral pH). EOF can be switched on/off easily through switching of the voltages that establish the electric field. Similarly, switching of flows from and to different liquid lines is possible without mechanical valves, allowing complex sample and reagent manipulation. This *electrokinetic valving* concept has been exploited for the injection of minimized sample fractions in capillary electrophoresis chips to separate DNA fragments (Harrison et al., 1993). The liquid velocity v_{EOF} corresponding to the EOF can be described with an equation similar to that for electrophoresis, but with a different mobility, μ_{EOF}:

$$v_{EOF} = \mu_{EOF}E \tag{2}$$

where

$$\mu_{EOF} = -\frac{\varepsilon\zeta}{4\pi\eta} \tag{3}$$

in which η and ε are the viscosity and the dielectric constant of the liquid, respectively, and ζ is the zeta potential, that is the electric potential that arises at a defined position in the double layer. This potential strongly depends on pH and wall material. Typically, for an aqueous solution with a salt concentration of a few mM, the electroosmotic mobility in a Pyrex glass tube at pH 7 is ~$4.8 \times 10^{-4}\,cm^2\,V^{-1}\,s^{-1}$. For typical electric field values of a few hundred volts per centimeter, a linear liquid velocity of a few millimeters per second will be achieved.

An important advantage of the use of EOF to pump liquids in a microchannel network is that the velocity over the microchannel cross section is constant, in contrast to pressure-driven (Poisseuille) flow, which exhibits a parabolic velocity profile. EOF-based microreactors therefore are nearly ideal plug-flow reactors, with corresponding narrow residence time distribution, which improves reaction selectivity.

Although for separation methods like capillary electrophoresis EOF is often deliberately suppressed by specific wall coatings, in most practical cases, the species which are dissolved in an aqueous solution and which are relevant for the desired synthesis will be charged and therefore they will experience both an electrophoretic (of the dissolved species themselves) as well as an electroosmotic (of the solvent) driving force. Since the direction of EOF for aqueous solutions in a glass microchannel is toward the negative electrode, for a cation the electrophoretic and electroosmotic forces add up to a larger positive value, while for an anion the two forces are opposite and lead to a retardation and possibly even to a negative velocity with respect to the direction of EOF and cations. In any case, species of different mobility become separated in different zones, at least if they originally were introduced as a plug into the main solvent stream, as is the case in capillary electrophoresis on a chip (Harrison et al., 1993). This concept has also been exploited in microreactors, in particular by researchers at Hull University in the United Kingdom (for literature references see below).

Spatial and temporal control of chemical reactions by electrokinetic principles relies on the possibility to direct reagents and products to or from selected points in a microchannel network, at specific times. An electrokinetic microfluidic network may be modeled as an electrical resistance network, using the well-known Kirchhoff voltage and current laws (Cummings et al., 2000; Fletcher et al., 1999, 2001, 2002; Qiao and Aluru, 2002), in which the flow rates are directly proportional to the electric currents. A pressure-driven microchannel network can be modeled with the same laws if they are rewritten in terms of hydraulic resistances and pressures and volume flows (Bula, 2009; Chatterjee and Aluru, 2005), but the electrokinetic case can be more complex if the local resistivity (and therewith the local electric field) changes due to electrophoretic separation and a phenomenon called *electromigration dispersion*. Such effects require the inclusion of the Kohlrausch's regulating function (Kohlrausch, 1897) in the model (Mikkers, 1999). We refer to the literature for more details about these phenomena and related effects like "stacking" which is often used in capillary electrophoresis to enhance separation. Deviations from the "ideal" uniform-resistivity-based network typically occur when a plug of an electrolyte with a resistivity largely different from the background electrolyte is injected in that background electrolyte. This would be analogous to a local variation in viscosity in pressure-driven flow, for example, for gas–liquid flow in a microchannel, or for injection of a highly viscous sample into a liquid chromatography column, leading to an instability called "viscous fingering" (De Malsche et al., 2009). That clean "switching" of flows in a simple Y-type microreactor inlet, using electrokinetic principles, is not as simple as it may seem, has recently been discussed in detail by MacInnes et al., who have shown that clean switching is difficult to achieve in practice, and that there is considerable contamination of each reagent supply channel with the other reagents (MacInnes et al., 2003). It has also been shown that the conductivity of the surface may play a significant role in electrokinetic flow principles (Fletcher et al., 2001) and that for extremely small channels, *nanochannels*, the situation becomes even more complex due to *electric double-layer overlap* (Sparreboom et al., 2009).

EOF has been applied as a pumping or mixing mechanism in microreactors (Fletcher et al., 2002). Mixing concepts include the introduction of two (or more, see Kohlheyer et al., 2005) parallel streams from a T- of Y-junction, where mixing at the low Reynolds numbers achieved occurs principally by interdiffusion of the two streams. This is a relatively slow process which may take tens of seconds to complete. Faster mixing can be achieved by injection of a sample of a specific composition via, for example, a double T-injector into a stream of liquid with a different composition.

Mixing speed is controlled by the width of the injected plug, and will be much faster than in the parallel-stream case (Fletcher et al., 2002).

Electrophoretic concepts have also been used for mixing purposes, for cases where the two reagents to be mixed have a different mobility, for example, applied in Wittig chemistry (Skelton et al., 2001a). Using electrophoretic principles to generate controlled concentration gradients of reagent streams, again in a Wittig synthesis, it was found possible to control the *cis* (Z) to *trans* (E) isomeric ratio in the product in the range 0.57–5.21, compared to a traditional batch method which gave a Z/E ratio in the range 2.8–3.0 (Skelton et al., 2001b). Several other examples of synthetic reactions carried out in a microreactor under electroosmotic or electrophoretic control, or both, have been reviewed by Fletcher et al. (2002). It was also demonstrated that similar principles can be applied with organic solvents, as long as they have sufficient polarity (Salimi-Moosavi et al., 1997). Watts et al. have shown that peptides may be prepared in quantitative conversion in a microreactor under electrokinetic control, with an increase in reaction efficiency compared with the traditional batch method. This increase was found to be due to an electrochemical effect (Watts et al., 2004). Finally, one specific other application of differential electrophoretic mobility is the selective control of product detection times in capillary electrophoresis in a method called *electrophoretically mediated microanalysis* (Bao and Regnier, 1992; Burke and Regnier, 2001).

In a previous section the concept of (simulated) moving bed chromatography was briefly discussed. Electrophoresis shows a resemblance to true moving bed chromatography, because negative and positive ions become separated and move relative to one another in opposite directions in an electric field. For species which have a charge of the same sign, a properly directed EOF, which can be tuned by changing the surface charge via an external electrode (Culbertson and Jorgenson, 1999; Schasfoort et al., 1999), via EOF modifiers or dynamic wall coating (Kaniansky et al., 1999; Melanson et al., 2001) or via a permanent surface coating (Revermann et al., 2008), or a well-tuned co- or countercurrent pressure-driven flow (Culbertson and Jorgenson, 1994) may establish the correct relative velocities for an electrophoretic moving-bed-like process. In a number of publications Thome and Ivory have described the continuous fractionation of enantiomer pairs using such an electrophoretic analog of moving bed chromatography (Thome and Ivory, 2002, 2006, 2007), using an industrial scale vortex-stabilized electrophoresis instrument. The concept is shown in Figure 24 (Thome and Ivory, 2007). It would be ideally suited for downscaling, as is the case for all electrophoretic principles, and similarly ideal for a downstream work-up method coupled to a (electrokinetically controlled) microreactor, but as far as we know this has not been attempted yet.

The methods described above all deal with direct current electroosmosis and electrophoresis. If electroosmosis is used with a time-periodic

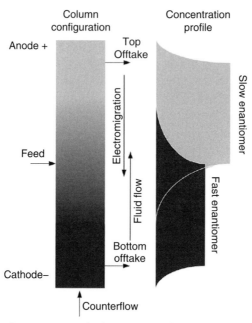

Figure 24 Electrophoretic moving bed separation of enantiomers. The downward movement of the enantiomers by electromigration is counteracted by a fluid flow upward. The fast enantiomer still has a net migration to the bottom outlet, while the slower enantiomer is pushed toward the top outlet (reprinted from Thome and Ivory, 2007, with permission from Elsevier).

electric field over a liquid volume, special effects can be generated. The first obvious effect is enhanced species mixing. Mixing is always an issue in microfluidic reactors, and electric fields in liquids may assist in this process in several ways. Above we have already discussed the application of electrophoresis to mix charged species without actual liquid movement. EOF induced by an unsteady electric field was experimentally shown to improve mixing (Oddy et al., 2001; Qian and Bau, 2002). The difference between the approaches of Oddy et al. and Qian and Bau is that in the first a flow instability, observed in sinusoidally oscillating, electroosmotic channel flow is used to generate stretching and folding of stream lines, while in the second periodic alternations of local ζ potentials is applied to induce chaotic advection (see also Chang and Yang, 2009). Spatial and temporal control of the ζ potential can be achieved by imposing an electric field perpendicular to the wall liquid, using electrodes embedded in the insulator wall (Schasfoort et al., 1999; van der Wouden et al., 2005).

It has also been demonstrated that nonuniform AC electric fields generated by neighboring coplanar microelectrodes produce a steady fluid flow in electrolytes (see Green et al., 2002, and refs. therein). The liquid moves from high-field strength regions on electrode edges to the surface of the electrodes, with the highest velocity found at the edge. The velocity depends on frequency and amplitude of the applied electric field, goes to zero at high- and low-frequency limits and has a maximum at a frequency that depends on the electrolyte conductivity (Green et al., 2000). Microfluidic devices with incorporated arrays of nonuniformly sized embedded electrodes, subjected to an AC field, were shown to generate a bulk fluid motion (Brown, 2000; Studer et al., 2002). The mechanism responsible for the liquid flow is the interaction of the tangential component of the electric field and the induced charge in the diffuse double layer on the electrode surface (Green et al., 2002), and will work best if an asymmetry is created in the coplanar electrode pattern, see Figure 25 (Brown, 2000). Typical liquid velocities that can be achieved with the method are 1 mm per second, at typical voltages of 1–10 V (rms).

4.2 Positioning and trapping of particles and molecules

AC electroosmosis as discussed in the previous section has not been used much as a pumping mechanism in microreactors, but the effect has been shown to be useful in (bio)particle manipulation, for example, in the work on an electrokinetic bioprocessor used for concentrating cells and molecules in a microfluidic system (Wong et al., 2004). In this device, the flow field generated by AC electroosmosis transports particles to regions near the electrode surface, where electrophoretic and *dielectrophoretic* forces, which are effective in short range, take over, and trap the cells and molecules. The concentration of biological objects in a large range of sizes, including bacteria, λ-phage DNA, and single-stranded DNA fragments with a radius of gyration of 3 nm, was demonstrated.

DEP is based on a force exerted on a dielectric particle by a nonuniform electric field. The strength of the force depends on the electrical properties

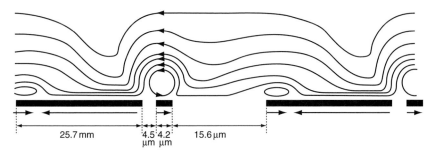

Figure 25 Predicted flow profile over asymmetric pairs of electrodes (from Brown, 2000; reprinted with permission. Copyright 2000 by the American Physical Society).

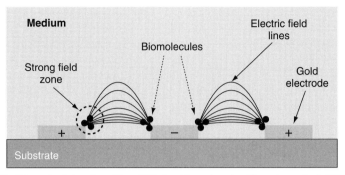

Figure 26 The principle of positive DEP, for biomolecules (from Luo et al., 2006; reprinted with permission from Elsevier).

of medium and particle, on particle shape and size, and on electric field frequency. DEP can be either *positive* (*p*-DEP), which is when the particle is attracted to a region of higher electric field because its permittivity exceeds that of the suspension medium (see Figure 26, Luo et al., 2006), or *negative* (n-DEP), attracting it to lower electric field if the permittivity of the medium is higher than that of the particle (Pohl, 1978). DEP is typically observed for particles with diameters of 1–1000 μm, but it is possible to apply it for smaller particles, and even biomolecules, as was shown in the work of Wong et al. (2004). For more accurate trapping of nanoscale objects like bovine serum albumin and antibody molecules (these have a size in the range of ~10–14 nm), Luo et al. have developed DEP with an array of nanoelectrodes with a width of 60 nm and a period of 380 nm, by which the EOF that disturbs trapping, is suppressed, and furthermore the required voltage is reduced by a factor of five in comparison with a microelectrode configuration (Luo et al., 2006). Although this has not been discussed in the work of Luo et al., the dimensions of the electrodes should not be very small in comparison with the microfluidic channel height (Markarian et al., 2003).

DEP can be used to create regular patterns of particles on a surface or in a microfluidic channel. For example, two-dimensional (2D) patterns of polystyrene latex microbeads were fabricated on glass substrates using n-DEP, and the line- and grid-patterned microparticles, which formed due to the repulsive force of this n-DEP, were covalently bound on the substrate via cross-linking agents (Suzuki et al., 2004). In this work, interdigitated microelectrodes were incorporated into a fluidic channel in order to direct the particles into very specific patterns.

Regular pattern formation in a particle deposit can occur spontaneously, through a self-assembly process that is based on the adhesive or repulsive interaction between particles and between particles and a surface. Electric fields may assist in pattern formation, via the forces between particles caused by their polarization in the DEP process, but also through electrophoretic (Trau et al., 1996) or electroosmotic transport (Solomentsev et al., 1997) of the medium or the particles in it along or toward a surface. Furthermore, to actively control the formation of colloidal crystals inside microchannels, a combination of electrocapillary forces and solvent evaporation was used (Shiu et al., 2004). Electrocapillary forces are capable of controlling fluid motion in three-dimensional (3D) structures, and the resulting well-ordered 3D patterns of particles could be very interesting for the field of catalysis in general, and catalytic microreactors in particular. Ordered packed beds would have a lower residence time distribution (in analogy with a lower theoretical plate height in chromatographic applications, see De Malsche et al., 2007) and therewith may give an improved reaction product selectivity.

Electroimmobilization of different enzymes in a microfluidic device has been described as a means to perform multistep reactions (Astorga-Wells et al., 2004). The procedure consists in first capturing target molecules by balancing opposing hydrodynamic (i.e., pressure-driven flow) and electric forces (i.e., EOF), after which a second medium, carrying enzymes or other reagents, is injected into the system and brought into contact with the target molecules to allow them to react. Disconnecting the electric field reverses immobilization and allows to collect the products at the outlet of the device. Examples of reactions that were carried out in this way are reduction, alkylation, and trypsin digestion of proteins. A similar concept has been used to separate peptides (Astorga-Wells et al., 2005).

CNTs and CNFs have already been discussed several times in this chapter. For applications in the microreactor field, they have become important as catalyst support because of their corrosion resistance to acids and bases, their high surface area, and the possibility to tune their surface chemistry (see De Jong and Geus, 2000 for a review). CVD is the most popular method of producing these carbon nanostructures, due to both the large-scale production feasibility and the possibility of controlling length, diameter, orientation, and growth locations of the structures. Typically, they grow from metal nanoparticles like Fe, Ni, Co, or their alloys, and then develop into 5–20 nm thick and a few to tens of micrometers long multiwalled nanotubes. Their electric behavior can vary from semiconducting to metallic.

However, a serious issue for device integration with CNTs is posed by the inability to control whether the tubes or fibers are semiconducting, semimetallic, or metallic. This aspect will also play a role if carbon nanostructures are used as a catalyst support. Except for a selective destruction of metallic tubes (Collins et al., 2001) an interesting method to separate metallic from semiconducting CNTs is the use of AC DEP. This is done by bringing a suspension of the tubes in contact with a microelectrode array. Due to the different dielectric constant of the species with respect to the solvent, an opposite movement of metallic and semiconducting tubes arises along the electric field gradient generated by the electrode array (Krupke et al., 2003). Subsequent work has shown that sorting improves with increasing AC frequency (Krupke et al., 2004). This dielectrophoretic separation concept has recently been implemented in a microfluidic setup, where the separation is facilitated by two parallel streams of liquid, where the metallic tubes are subjected to a significantly larger dielectrophoretic force perpendicular to the flow direction and drawn into the second stream, while the semiconducting tubes remain in the first stream (Shin et al., 2008). This approach, which is very similar to earlier work based on dielectrophoretic field-flow fractionation (Peng et al., 2006), significantly increases throughput. The final goal of this work is the fabrication of nanoelectronic devices, but the use of such concepts can become very valuable for manipulation and positioning of catalytic nanoparticles in future. For example, positive DEP was used to interface CNTs and catalytic Pd to realize a H_2 gas sensor (Suehiro et al., 2007). CNTs were either trapped alone, or simultaneously with Pd nanoparticles (giving CNTs modified with Pd nanoparticles), on Pd electrodes.

Recently reports have appeared in literature about the control of *single* nanoscale objects in liquids by electric fields. Control of micrometer-scale objects had been reported earlier already, but for smaller objects special care has to be taken to compensate for the Brownian motion of the particles. Cohen et al. (Cohen, 2005; Cohen and Moerner, 2005) created an electrophoretic trap which creates arbitrary 2D force fields for individual nanoscale objects in solution, couples fluorescence microscopy with digital particle tracking, and real-time feedback to generate a position-dependent electrophoretic force on a single nanoparticle. The control of position for a 20 nm particle was within a few microns. It is stated that positional control of a single fluorophore is fundamentally limited by the finite rate of photon detection. State of the art would allow the control of a typical fluorophore within 1 µm, if sufficient electric field could be generated, which would require an electrode pattern with an interelectrode gap on the order of 3 µm. This is indeed possible with current photolithographic techniques. The authors have demonstrated trapping and manipulation of single virus particles, lipid vesicles, and fluorescent semiconductor nanocrystals (Cohen and Moerner, 2006). This work could also become very relevant for the study of the chemical reactivity of single surface-modified beads, single catalyst particles, and single enzymes.

As a final demonstration of new technological achievements, it was recently also demonstrated that *gas molecules* can be trapped in an electric field on a chip (Meek et al., 2009). Manipulating a packet of ions in a vacuum is quite common practice in mass spectrometric instrumentation, where ions can be collected and analyzed in, for example, ion traps and collision cells, and their flow can be steered by electrostatic lenses (and magnetic fields) through time-of-flight detectors and other sophisticated ion separators and analyzers. In the work by Meek et al., though, CO molecules, loaded directly from a supersonic beam, were confined in tubular electric fields of 20 μm in diameter and centered 25 μm above a chip. These field traps move with the molecular beam at a velocity of several hundred $m s^{-1}$ and can hold molecules for a certain time, and subsequently release them off the chip for detection. The authors claim that this methodology is applicable to a wide variety of polar molecules, and enables "the creation of a gas-phase molecular laboratory on a chip" (Meek et al., 2009). Although these authors discuss their results mainly in the light of quantum optic and solid-state technology (quantum computing), one can foresee applications for chemical synthesis as well, if such microstructures can be reconfigured for the manipulation and recombination of small packages of different molecules, in combination with fast spectroscopic methods to study chemistry.

4.3 Electrowetting-on-dielectric

The use of droplets as microreactors has been discussed already many years ago (Pileni, 1993), but the combination with microfluidic networks which give control over individual droplets and their composition has brought upon a completely new field of research, which, since it works with discrete liquid packets, may also be referred to as *digital microfluidics*. Digital microfluidics shows a clear analogy with traditional benchtop protocols, and a wide range of established chemical protocols can seamlessly be transferred to a picoliter to nanoliter droplet format. *Electrowetting*, that is, the change in surface tension caused by an electric field (Berge, 1993), DEP, and immiscible liquids in pressure-driven flow are the three most commonly used principles used to generate and manipulate the droplets in a digital microfluidic device.

EWOD is used to control the flow of liquid droplets on a surface or in a channel and works both with droplets in air, with droplets of a liquid immiscible with the surrounding liquid (e.g., water–oil combinations), and even with air bubbles in a liquid (Zhao and Cho, 2007). Typically, in an electrowetting device, a voltage of 50–100 V is applied across a 1-μm-thick insulator, which causes the contact angle of an aqueous solution on a hydrophobic surface to decrease from ~115° to ~75°. But by using a very thin (70 nm) and high dielectric constant (~180) material, only 15 V are needed to give about the same contact angle change (Moon, 2002). The theory of EWOD have been discussed in detail in recent reviews (Mugele, 2009; Mugele and Baret, 2005), and for an overview of applications of droplet microfluidics, including EWOD, we refer to another recent review (Teh et al., 2008).

Here we want to highlight the use of EWOD in microfluidic networks or on surfaces with embedded electrodes as microreactors. Several research groups have worked on this concept, and each group has its own favorite design, operation regime or application field (Pollack et al., 2002; Taniguchi et al., 2002). EWOD-based mixers have been described that work either on the fast displacement of droplets on an array of electrodes and the accompanying internal flow in the droplet generated by the rolling movement (Paik et al., 2003), or on the periodic change of contact angle of the droplet (Mugele et al., 2006; Nichols and Gardeniers, 2007), which leads to stretching and relaxation of the droplet, and therewith to a sort of "shaking" effect that speeds up mixing. Although a fundamental explanation for the observed behavior is still lacking, it was found possible to mix an enzyme with its substrate in a time as short as 15 ms with EWOD at an optimal AC frequency of 750 Hz, in a study on the pre-steady-state enzyme kinetics of a specific phosphatase (Nichols and Gardeniers, 2007). The reaction mixture was then rapidly quenched with an acid by the same principle, and subsequently mixed with a matrix solution in order to prepare it for MALDI-TOF (Matrix Assisted Laser Desorption Ionization Time-of-flight) mass spectrometry, a method that, because of its off-line character, matches perfectly with digital microfluidics (Wheeler et al., 2004).

4.4 Special effects

In this section we would like to briefly describe a number of less developed and in some cases unexplained ideas and phenomena based on electric fields, which have been or may be used in combination with chemical (micro)reactors. They are posed here as possible suggestions for further research.

4.4.1 Electric wind

The first principle is that of *electric wind* (or ion wind, ionic wind, or coronal wind), the earliest report of which was already made in 1709 by Hauksbee (Robinson 1962). Since electric charge resides entirely on the external surface of a conductor and concentrates around sharp points and edges, the electric field on a point or edge is much higher than that on a flat or smooth surface. When this field exceeds a certain strength, known as the corona discharge inception voltage gradient, it ionizes the air close to the tip. The ionized air molecules have the same polarity as the tip, so that they are repelled and create an electric "wind" in a direction away from the tip.

One of the applications of electric wind is in aerodynamics, in which field it has been an important objective to modify airflow around an obstacle in order to reduce drag. The advantage of flow control by an electrohydrodynamic concept that directly converts electrical energy into mechanical energy is that no moving mechanical parts are involved and a very short response time, because of the electric control, is established (Magnier et al., 2007; Moreau, 2007). Another application is in spacecraft propulsion, where an *ion thruster* creates thrust by accelerating ions (Lerner, 2000). Although the thrust created in ion thrusters is quite small compared to chemical rockets, a high propellant efficiency is obtained because of the very limited propellant consumption. Electric wind also helps in the enhancement of heat transfer (Kalman and Sher, 2001) and was suggested and tested as a concept for cooling of microelectronic circuits (Go et al., 2007) where thin-film electrode patterns of a stable material like diamond with gaps of several micrometers were shown to be particularly useful (Go et al., 2009). Experimental work demonstrated a more than twofold enhancement of the local heat transfer coefficient with ionization combined with typical externally forced flow conditions (Go et al., 2007). This concept, and several other principles of *microscale thermal transport* such as pool boiling heat transfer enhancement and convective flow boiling in microchannels, which may become relevant for implementation in microreaction technology as well, have recently been reviewed (Garimella and Liu, 2009).

A direct application to chemical process technology of the principle of electric wind is in *electrostatic precipitators* (Leonard et al.,1983) and *electrocyclones* for size separation of particles in powder technology (Nenu et al., 2009). Electrostatic precipitators applied to exhaust gas cleaning have recently been reviewed (Jaworek et al., 2007). A particularly interesting development is that of a small electrocyclone with a diameter of 75 mm (Shrimpton and Crane, 2001). With this device it was shown that the separation quality of the smallest size particles with a diameter below 38 μm doubled upon application of the electric wind. Later experiments performed with submicron silica particles demonstrated that classification of such particles is possible by use of an electrical hydrocyclone (Nenu et al., 2009).

4.4.2 Electric swing adsorption

A very recent development, with the objective of reusing greenhouse gases like CO_2, is *electric swing adsorption* (Moon and Shim, 2006). As an alternative to pressure swing adsorption, which utilizes a vacuum, or thermal swing adsorption, which applies a hot inert gas, to recover a gas that was separated from a mixture by selective adsorption on a specially selected material (like a zeolite), electric swing adsorption utilizes a direct electrical current to heat the material, so that the gas desorbs quicker and with lower energy consumption. The feasibility of the concept for capturing CO_2 from flue gases with 3.5% CO_2 from natural gas power stations was studied, and it was found that an adsorbent consisting for 70% of zeolite and for 30% of a conducting binder material can give a concentrated stream with 80% CO_2 with an energy consumption of 2.04 GJ per ton of CO_2 (Grande et al., 2009). This work indicates that electric swing adsorption may become an interesting CO_2 capture technology.

4.4.3 Pulsed electric fields

Pulsed electric fields are used in food preservation to inhibit microorganisms in foods without significant loss of flavor, color, taste, and nutrients. The treatment generates a very short (microseconds) but high electric current pulse through the food. The topic has recently been reviewed, including the fundamental mechanisms of food preservation (Min et al., 2007). The main mechanisms for microbial inhibition are structural damaging of cells, in particular cell membranes, due to a number of mechanisms related to cell stress and fatigue, and inactivation of enzymes due to association or dissociation of functional groups, movement of charged chains, and changes in the alignment of helices.

A peculiar recent report is that in which the application of an AC high voltage to accelerate wine aging is discussed (Zeng et al., 2008). The optimum treatment was found to be with an electric field of 600 V cm^{-1} for 3 min. Several analysis techniques were used to clarify the differences between treated and untreated wine samples, and it was found that the amount of higher alcohols and aldehydes in volatile compounds decreased significantly while the amount of some of the esters and free amino acids slightly increased. A mechanism that may be responsible for the formation of the free amino acids is protein degradation, but otherwise no explanations have been given for the effect of the AC electric field.

4.4.4 Electrospray

Electrospraying is a concept in which an electric field is applied between a liquid outlet and an object at some distance from the outlet. The liquid becomes dispersed and develops into a very fine aerosol because droplets break up due to a Rayleigh instability, that is, the phenomenon that a thin beam of a liquid in a gas becomes unstable at a certain length and breaks up into droplets, which effect is enhanced by an electric field on the liquid (Rayleigh, 1882). A related concept that is used very frequently to generate ions that can be separated in a mass spectrometer is *electrospray ionization* (ESI). Here a high voltage is applied to a liquid supplied through a nozzle (e.g., a small glass capillary). The liquid forms a so-called Taylor cone, at which apex droplets break up into a fine spray with smaller and smaller droplets because of radial dispersion due to Coulomb repulsion in combination with solvent evaporation, until finally all the solvent is evaporated and all the solutes have become ionized and can enter the mass spectrometer inlet to become separated and detected. ESI has been reviewed many times in literature, and has become a very valuable concept for coupling microreactors and other (analytical) microdevices to a mass spectrometer (Le Gac and van den Berg, 2009).

Electrospray has been used for many different applications, such as the deposition of paints and coatings on metal surfaces and the deposition of metal nanoparticles and biomolecules on biosensor surfaces, and in a miniaturized version also as a propulsion mechanism in microsatellites (see also the section on electric wind). One particularly interesting application is in *fuel atomization*, that is, a finer fuel aerosol and atomization will give a higher combustion efficiency and less pollutant emission, which is caused by the effect that finer droplets increase the total surface area on which combustion can start (Lehr and Hiller, 1993).

There has been a claim of another useful effect of electric fields on fuels, that is, an electric charge on a fuel tube is said to reduce fuel viscosity, so that fuel injectors will create smaller droplets, which enables cleaner and more efficient combustion. The principle is that molecules in the fuel become charged and aggregate, reducing their overall surface area, therewith decreasing the viscosity (Tao et al., 2008). The paper by Tao et al. received serious criticism, where the criticism is that, although the application of an electric field might indeed reduce fuel viscosity and lead to finer atomization, there is no proof that finer atomization would improve combustion efficiency by 20%, while it is also stated that the earlier "claims and conclusions violate the first law of thermodynamics" (Gulder, 2009). This, as might be expected, is rebutted by the authors of the first paper (Tao et al., 2009). Nevertheless, this idea may become a topic with a high impact, and surely deserves further research.

5. ELECTRONIC CONTROL OF REACTIONS AT SURFACES

5.1 Adsorption–desorption controlled by electrical fields

Electric fields at field emitter tips, as discussed before, are typically on the order of $10 \, V \, nm^{-1}$. This is in the same range as the electrostatic fields that are present in zeolite cages (see below) and at the interface of electrode and electrolyte interface. Since these fields are all of the same order as the fields inside atoms and molecules, they are strong enough to induce the rearrangement of electronic orbitals of atoms and molecules. It is therefore expected that it should be feasible to stimulate chemistry with such electric fields.

In this section we are particularly interested in *field-induced adsorption/ desorption* (Kreuzer, 2004). In the review by Kreuzer, electric field effects are classified either as physical, for fields below $10 \, V \, nm^{-1}$ which mainly give rise to polarization, or as chemical, for fields above $10\,V\,nm^{-1}$, which give rise to such a level of distortion of electronic orbitals of atoms or molecules that new bonds may form. Far from the surface and in the absence of an electric field, the atomic orbitals of two atoms hybridize into two molecular orbitals, a lower bonding and a higher (empty) antibonding orbital. As the molecule approaches the metal surface, additional hybridization with the conduction electrons occurs, leading to shifts and broadening of these orbitals, so that the antibonding orbital becomes partially occupied, resulting in bonding to the surface. An electric field pointing away from the surface adds a potential energy for electrons outside the metal, which raises the atomic levels of atoms A and B, resulting in a rearrangement of the molecular orbitals. For this situation, the antibonding orbital becomes empty again, and the surface bond probably weakens. At a field strength where the bonding orbital is lifted above the metal Fermi level, this orbital will also drain, and field-induced dissociation will occur.

This principle has been applied in *field ion microscopy* (FIM), in which a (pulsed) electric field generated by a sharp metal tip allows the investigation of the field effect on chemical reactions (Block, 1963). Typically, the potential at the tip is *positive* and generates a field close to it of $\sim 10\,V\,nm^{-1}$, the chamber in which the tip is applied is at high vacuum, and the tip is cooled to temperatures of ca. 20–80 K. Quite similar to FIM is *field emission microscopy* (FEM), which also applies a sharp metal tip but at a large *negative* potential, giving an electric field near the tip on the order of $\sim 10\,V\,nm^{-1}$, high enough for field emission of electrons to occur. Both these techniques are valuable for studies of the interaction of gases with catalyst surfaces, as has been shown recently in an investigation of the interaction of carbon monoxide and oxygen with gold (McEwen and Gaspard, 2006). In this work a field-dependent kinetic model based on the Langmuir–Hinshelwood mechanism was constructed to show that dissociative adsorption of oxygen on gold occurs only below a negative critical electric field value, while binding of CO on gold is enhanced for positive field values. Although the experimental conditions are far from the ones in practical catalytic chemistry (on the surface of a metallic tip the electric field varies according to the local radius of curvature, leading to spatial differences in adsorption and desorption of molecules, while also pressure and temperature are significantly different), this research can still give important clues about the surface processes occurring on a catalytic particle (like a Au nanoparticle with a diameter of a few nm) on a specific support (see Figure 27). The charged metal tips not only mimic the well-known (electronic) metal–support interaction in catalysis (Yoshitake and Iwasawa, 1992), but also modify the pressure of the gases close to the surface of the catalyst particle due to the polarizability and the electric dipole of gas molecules. Using the equilibrium statistical mechanics of diatomic molecules in an electric field, the effective potential of the molecules at 300 K can be approximated by a quadratic

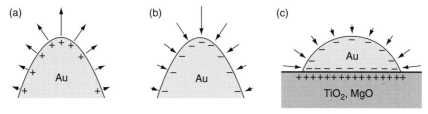

Figure 27   (a) positively charged gold tip in FIM, (b) negatively charged gold tip in FEM, and (c) negatively charged gold cluster on an oxide solid substrate (from McEwen and Gaspard, 2006; reprinted with permission. Copyright 2006, American Institute of Physics).

function of the field, and from that the dependence of the pressure of a specific molecule on the electric field F can be approximated by a function of the general form:

$$P(F) \cong P(0)\exp\{\beta u_{\mathrm{eff}}\} \qquad (4)$$

where $\beta = (k_{\mathrm{B}}T)^{-1}$, with k_{B} Boltzmann's constant, βu_{eff} the mentioned effective potential of the molecule, and $P(0)$ the pressure at a position where the electric field can be supposed to vanish. In the cited work of McEwen and Gaspard, the effective potentials of O_2 and CO are 0.022 and 0.028 $(\mathrm{V\,nm^{-1}})^{-2}\,F^2$, respectively, which at a tip electric field of $10\,\mathrm{V\,nm^{-1}}$ gives pressure enhancements of factors 9.5 and 17, respectively (McEwen and Gaspard, 2006).

Above we already mentioned that the electric fields achievable at a metal tip to which a voltage is applied can be in the same range as the electrostatic fields that are present in zeolite cages, for example, the electrical fields present on unshielded alkali ions in zeolite Y are estimated to be on the order of 3–$10\,\mathrm{V\,nm^{-1}}$ (Blatter et al., 1994). Such large fields stabilize charge-transfer states of properly oriented colliding hydrocarbon and oxygen pairs, so that reactions between them can be initiated at the low energies of visible photons or even by mild heating (Frei, 2006). Similarly, electrostatic fields play a role in the ability of a *metal-organic framework* (MOF) material to separate gas mixtures (Yang and Zhong, 2006). The selective gas adsorption in an MOF is related to both the pore size and the interactions between adsorbed molecules and the pore walls. If the MOF has surfaces with high electrostatic fields or field gradients, the adsorption of molecules with a high dipole or quadrupole moment, respectively, will be enhanced, for example, in the MOF of type HKUST-1 (Li et al., 2009).

A question that arises is whether one can modify the electrostatic fields (or their gradients) around active sites in zeolite cages or around the pores in an MOF, by applying an *external field*. As far as we know experiments of this kind have not been reported, but a possible experimental configuration to do so is a capacitor in which the zeolite is sandwiched between two charged electrodes of a conductor (which should preferably be catalytically inactive, excluding most of the noble metals). A way to analyze the local intracrystalline electric field gradients in the zeolite and their possible modification by external electric fields is NMR of quadrupole spins, like those of adsorbed ^{131}Xe (Millot et al., 2001). It is doubtful whether the zeolite or MOF in the mentioned capacitor structure can withstand the electric fields required to get a significant effect on the catalytic activity. For example, the breakdown field strength of dense silicon dioxide, one of the most reliable insulators available in microelectronics, is $\sim 1\,V\,nm^{-1}$, and breakdown of zeolites of MOFs will most likely occur at lower fields than this. This is why in our laboratory we are studying configurations based on insulated nanoneedles, at the tips of which the electric field will be significantly enhanced.

The field of fine and pharmaceutical chemistry, where the desired production amounts are in kilograms rather than in tons, would benefit considerably from the development of versatile process chemistries, carried out in relatively small generic production units that can be tuned to a range of different products. An example of a versatile "microreactor" is a microorganism of which the product can be tuned by engineered genetic changes that alter the metabolic processes of the organism in such a way that it produces a desired chemical. The key actors in this are the different enzymes which are expressed by the microorganism and which perform the required biocatalysis on very specific feed molecules to deliver the desired product molecules with extremely high selectivity, for example, as a single enantiomer. Enzymes in living cells are often not present free in solution but immobilized in the lipid membranes of cell organelles or of the cell outer membrane. It is difficult to replicate membrane immobilization outside of a cell, in a technical device, which is why for industrially relevant enzymes immobilization strategies have been developed based either on support binding (usually involving covalent bonds, for example, to a resin, a biopolymer or a mesoporous silica), entrapment (in a gel lattice), or cross-linking of enzyme aggregates or crystals to prepare carrierless particles (Sheldon, 2007).

For reasons of fouling or deactivation, it may be desirable to be able to replace immobilized enzymes in time, and for this a programmable way of adsorption and release of enzymes would be very welcome. An example of this is the use of a 4 nm thin polymer film that can be thermally switched between a hydrophilic (swollen) state at 20 °C and a more hydrophobic protein-adsorbing (collapsed) state at 48 °C, integrated into a micro hotplate with fast heating options so that a protein monolayer can be adsorbed and released within 1 s (Huber et al., 2003).

The application of heat to enzymes is not always desired; furthermore, it may turn out to be difficult to localize the heat. Electric fields have an advantage in this respect, since they can be locally applied by design of electrode configurations. Besides the concepts which were already mentioned earlier in which electric fields are applied to transport molecules or particles to a desired location, or trap them at a specific location, there are also ways to control the adsorption of molecules from solution on a specific surface.

Two principles exist by which electric fields can control molecular adsorption on a surface. The first is based on a change in surface wetting properties, as was discussed in a previous section, where EWOD was applied to switch from a hydrophilic to a hydrophobic state and *vice versa*. The second relies on changing the surface charge density, for example, on a metal electrode in contact with an electrolyte, or changing the ζ potential, by applying an external electric field on electrodes embedded in an insulator surface on which one desires to change the affinity. Experiments involving the adsorption of organic molecules at a solid surface have shown strong capacitance changes in the electric system in which the solid surface was included, which changes are associated with phase transitions in the adsorbed layer between gas-like layers, disordered and ordered physisorbed layers, and chemisorbed layers (Van Krieken and Buess-Herman, 1999). The concept of changing the ζ potential on a surface arranged in a field-effect transistor (FET) configuration by a *gate potential* was discussed in the section on EOF as a method to control that flow, but it should also be possible to control in similar manner the (reversible) adsorption of molecules, especially of molecules which have a charge or a dipole. The effect is very sensitive to the pH of the electrolyte with which the surface is in contact, and typically the largest effect of the *field-effect flow control* principle occurs close to the pK value of the surface of interest (Schasfoort et al., 1999; van der Wouden et al., 2005). This may prohibit the use of the concept for enzyme immobilization because also the mobility and solubility of proteins is very much pH dependent. Attempts in our laboratory (Nichols and Gardeniers, 2006) to use field-effect control of protein adsorption were unsuccessful, which may have been due to the problem of balancing the charge on the surface (controlled by the pH, ionic strength, and both the electric fields along and perpendicular to the surface) with the charge on the protein (controlled mainly by pH, and the choice of protein). Furthermore, in a theoretical study on electrostatic contributions to the entropy and energy of protein adsorption (Roth et al., 1998) it was shown that the attractive free energy is very strong at short range. The protein was modeled as a colloidal particle, with an electric double layer that starts to overlap with the electric double layer of the surface at short range, leading to a significant entropic effect due to ion transport from the double layers to the bulk of the solution. This (positive) entropic effect is said to dominate the free energy of adsorption. However, this effect very much depends on the matching of charges on the protein and on the wall, and in some cases this may even lead to a repulsive force between protein and wall (Roth et al., 1998). Control seems therefore to be difficult or at least unpredictable. In addition, there is also always a risk that, due to the distortion of the ionic mantle around a protein molecule by a strong electric field, particularly at high protein concentrations, the protein solution becomes instable and starts to aggregate or precipitate, like it is the case for sols that may become destabilized by a change in ionic strength.

Probably, a combination of electrostatic effects (tuning the charge on a surface) and wetting effects is a better concept to control *reversible* and *nonspecific* adsorption of proteins. Proteins and other biomolecules in aqueous solutions tend to nonspecifically (i.e., not via their active bonding site) adsorb onto hydrophobic surfaces, but as mentioned above they can also become adsorbed through electrostatic attractions. It was shown that protein adsorption in an EWOD configuration could be minimized by limiting the time during which no potential is applied (which is the nonwetting i.e., hydrophobic state) and through choice of solution pH and electrode polarity (Yoon and Garrell, 2003). In another study, the adsorption kinetics of human serum albumin and horse heart cytochrome c under the influence of an electric field, from aqueous solution onto an indium tin oxide (ITO) electrode were investigated, with the aim of producing a surface-immobilized protein layer of tailored structural properties (Brusatori et al., 2003). With the aid of optical waveguide light-mode spectroscopy, it was found that adsorption rates changed according to the charge of the proteins in relation to the applied potential, but depended on the chosen buffer. The rate of adsorption at high surface density increased with the voltage for both studied proteins, which effect was more pronounced in water than in a HEPES buffer solution. This effect was attributed to contact between electrode and protein patches of complementary charge, leading to more oriented and more efficiently packed (multi)layer formation. It is important also to note that the metalloprotein azurin, entrapped on a silicon dioxide surface in-between metal electrodes of $1\,\mu m$ width and with a gap of $1\,\mu m$, keeps its native configuration up to electric fields of 10^{-3}–$10^{-2}\,V\,nm^{-1}$ (Pompa et al., 2005).

As a final topic in this section we would like to discuss a study aimed at the development of protein microarrays (i.e., planar substrates with regular spots with immobilized proteins of which the interaction with specific molecular probes or cells in solution is tested), in which an electrochemical switching strategy was implemented in order to modify surface properties (Tang et al., 2006). An ITO-based microelectrode array was uniformly coated with a protein-resistant polymer, poly-(L-lysine)-grafted-poly(ethylene glycol) (PLL-g-PEG). The electrical field induced by microelectrodes electrochemically removes the PLL-g-PEG adlayer from conductive areas, while the surrounding insulating surface is unaffected and remains protein resistant. The unprotected microelectrodes are then surface functionalized with the desired biomolecules such as DNA, proteins, or lipid vesicles with membrane proteins.

5.2 Control of the activity of adsorbed molecules

Besides the control of the adsorption and desorption of molecules on a surface, an exciting topic is tuning of the *orientation* and the *activity of adsorbed molecules* by an electric field. For organic molecular layers adsorbed on electrodes, it was found that after the initial adsorption, a sufficiently high (positive) potential leads to orientation of the molecules, where the molecular dipoles align with the electric field so that adsorbate–adsorbate interactions are increased (Han et al., 2004; Pronkin and Wandlowski, 2003). For the zwitterionic molecule *p*-aminobenzoic acid (PABA) on an Ag(111) electrode surface also a potential-dependent orientation effect was observed, using infrared–visible sum frequency generation spectroscopy and electrochemical capacitance and CV measurements (Schultz and Gewirth, 2005). It was found that PABA switches orientation with the charge on the electrode surface, orienting one way for potentials above the potential of zero charge, and the other way below that potential, an effect that depends on pH.

While in the mentioned work on PABA on a charged silver electrode also the adsorption degree of the molecules was controlled by surface charge, Lahann et al. showed a concept in which molecules immobilized as a low-density self-assembled monolayer on a gold surface were electrically stimulated to undergo conformational transitions between a hydrophilic and a hydrophobic state (Lahann et al., 2003). Such a surface wetting switch may then be used to immobilize, for example, enzymes, as was discussed in the previous section. This is an example of switching both the orientation and the activity of adsorbed molecules.

Adsorbing proteins in the correct orientation on a surface is very relevant for immunoassays, in which antibodies are immobilized on a surface and used to detect the presence and concentration of specific antigens in a biological sample, but also for enzymatic microreactors, where the orientation (and folding state) of the enzyme determines its biocatalytic activity. Many reports have appeared in literature on the orientation of proteins on charged surfaces, both on theoretical and experimental work. Theoretical studies are based on two different types of models (Sheng et al., 2002): molecular models based on Molecular Dynamics or Monte Carlo simulations, which are very detailed at the atomic level but also very demanding in terms of computing power, and continuum models which treat proteins as colloidal particles with a (mostly uniform) charge distribution over its spherical surface and the solvent as a continuum dielectric medium. A typical example of the latter is the Derjaguin–Landau–Verwey–Overbeek theory, which includes screened Coulombic repulsions and van der Waals attractions between proteins and surfaces. This theory describes most of the essential features of protein adsorption (see also the work of Roth et al. described in the previous section), but due to the assumption of a spherical protein with a charged shell inherently is unable to describe orientation effects. Therefore a compromise is often taken between the two types of models, by considering, for example, an immunoglobulin protein as a Y-shaped object with located charge centers on its branches and Monte Carlo modeling it at the mesoscopic scale (Sheng et al., 2002). In accordance with experimental data on oriented adsorption, Sheng et al. have shown that orientation of the adsorbed protein is based on a balance of van der Waals and electrostatic interactions. Van der Waals attraction leads to molecules lying flat on the surface, while in electrostatically dominated adsorption the orientation is determined by the dipole of the molecule which generally leads to alignment perpendicular to the surface. Experimental methods to probe protein orientation on charged surfaces are SIMS (Secondary Ion Mass Spectrometry) (Wang et al., 2004) and, more directly, atomic force microscopy (Wang et al., 2006). The latter work is particularly interesting because it reports that carbonic anhydrase is oriented with the majority of its active sites facing upward on a positively charged surface, and downward on a negatively charged surface, which shows that in principle the activity (or at least the accessibility of active sites on the enzyme) is controllable by the charge on the surface, which can be controlled by setting a specific voltage to the electrode.

An interesting effect of external electric fields on the selectivity of a homogeneous catalyst (an iron-oxo porphyrin compound) has recently been theoretically predicted (Shaik et al., 2004). The effect consists in changing the electronic structure of the catalyst in a field-direction-dependent manner. In doing so, the catalyst can be tuned either to epoxidation or to hydroxylation of propene. The electric field should be aligned along the SH–Fe–O axis of the molecule, which is in the direction perpendicular to the porphyrin ring. If the field is positive (i.e., pointing from S to O) the catalyst will acquire a thiolate radical character, therewith favoring the hydroxylation process by $6-10 \, \text{kcal mol}^{-1}$, while a negative field, turning the compound into a porphine radical cationic species, favors epoxidation by $2-6 \, \text{kcal mol}^{-1}$. Although a very interesting concept, experimental proof for the validity of this work has not yet been presented. An important issue to solve for practical evaluation of this idea is that one needs to fix the molecule in some way with respect to the electric field, for example, by immobilizing it on a surface, but in such a way that the porphyrin ring is still accessible from both sides and also with such a link to the surface that the activity is not affected. This is not an easy problem to solve, and possibly the use of crystalline mesoporous materials like zeolites or MOFs may help, if a way is found to incorporate porphyrin-based catalysts in these materials in an ordered manner. We are not aware of any research in this direction.

Because of the current great interest in alternative energy in general and artificial photosynthesis in particular, and possible future developments of microreactors designed to perform photosynthesis, the following work is also worth mentioning. Protein-mediated electron transfer plays a key role in photosynthetic or metabolic reactions in living organisms, where such reactions generate electrons with a relatively low reduction potential. It has been established that several electron transfer reactions are rate limited by conformational changes at interfaces, for example, at a protein–protein interface or across an electrode–protein interface. In a study on cytochrome c which was electrostatically bound on Ag electrodes coated with a self-assembled monolayer of carboxyl-terminated alkylthiols the effect of the electric field strength on the activation energy of interfacial redox process was determined (Murgida and Hildebrandt, 2002). By varying the alkyl chain length and therewith the protein–electrode distance, the field strength was controlled. A high electric field increases the activation barrier for the structural reorganization of the protein, and rearrangement of the hydrogen bond network becomes the rate-limiting process in the interfacial redox process. It is concluded that this *electric-field-induced change in the activation barrier* which controls interfacial electron transfer dynamics in the case of cytochrome c might represent the modulation of biological charge-transfer dynamics at cell membranes. In a more recent publication (Kranich et al., 2008) two-color, time-resolved, surface-enhanced resonance Raman spectroelectrochemistry was used to monitor simultaneously and in real time the structure, electron-transfer kinetics, and configurational fluctuations of cytochrome c electrostatically adsorbed to electrodes in the same manner as described above. It is found that the interfacial electric field controls protein dynamics, which in turn control the overall electron transfer.

A controlled modification of the rate and selectivity of surface reactions on heterogeneous metal or metal oxide catalysts is a well-studied topic. Dopants and metal–support interactions have frequently been applied to improve catalytic performance. Studies on the *electric* control of catalytic activity, in which reactants were fed over a catalyst interfaced with O^{2-}-, Na^+-, or H^+-conducting solid electrolytes like yttrium-stabilized zirconia (or electronic–ionic conducting supports like TiO_2 and CeO_2), have led to the discovery of *non-Faradaic electrochemical modification of catalytic activity* (NEMCA, Stoukides and Vayenas, 1981), in which catalytic activity and selectivity were both found to depend strongly on the electric potential of the catalyst potential, with an increase in catalytic rate exceeding the rate expected on the basis of Faradaic ion flux by up to five orders of magnitude (Vayenas et al., 1990). In a recent review (Vayenas and Kout-sodontis, 2008), the current understanding of the mechanism and practical implications of NEMCA (or *electrochemical promotion*) have been discussed. It is important to note that short-circuiting catalyst and counter electrode (i.e., without external power source) is sufficient to induce NEMCA, which in that case is controlled by the spontaneous potential difference between catalyst and counter electrode due to their different activity for the catalytic reaction. Furthermore, if the support has mixed electronic and ionic conductivity (effectively giving an internal short circuit), like TiO_2, then NEMCA is induced even without external short circuit and therewith the system shows quite some similarity with the well-known metal–support interaction (Ertl et al.,1997; see also Figure 27c).

All the available experimental and theoretical work performed on NEMCA leads to the conclusion that electrochemical promotion is caused by electrocatalytic introduction of promoting species like O^{2-} or Na^+ from the solid electrolyte to the catalyst/gas interface where a double layer is formed, of which density and internal electric field vary with the applied potential. The latter affects the work function at the surface and therewith the bond strength of adsorbing reactants and intermediates. This causes the dramatic and reversible modification in catalytic rate (Vayenas and Koutsodontis, 2008; see Figure 28).

The mechanism behind the NEMCA effect (which is active in gases and at relatively high temperatures, of a few hundred degrees, where solid electrolytes have a significant conductivity) suggests that there may be another way of controlling activity on a catalyst particle, particularly in the liquid phase: the modification of the metal–support interaction by means of external fields. On may call this a *catalytic field-effect transistor* or *cat-FET* (although it is not used here as a transistor element). The hypothesis is that a field effect similar to that used in a transistor, may be used to control electronic depletion of a catalytic particle on an insulating support

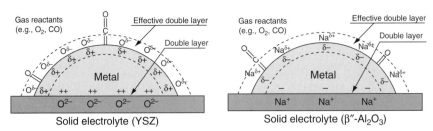

Figure 28 Metal electrode on O^{2-}-conducting (left) and Na^+-conducting (right) solid electrolyte. The figure depicts the metal–electrolyte double layer at the metal–gas interface due to electric potential-controlled ion migration, as well as its interaction with adsorbed reactants during CO oxidation (from Vayenas and Koutsodontis, 2008; reprinted with permission. Copyright 2008, American Institute of Physics).

material, and in that sense it relates to the metal–support interaction discussed earlier. The configuration envisioned is that of Figure 27c, but with a counter electrode at some distance from the metal particle, and an electrode on the backside of the support, to create an electric field over the particle–support interface. The idea relies on literature reports stating that the metal–support interaction on oxides involves a change in the electronic properties of the metal particle, correlated to the electron richness of the oxygen atoms in the support (Mojet et al., 1999), which is determined primarily by the ionic character of the cations in the oxide. In basic supports with alkaline cations, oxygen atoms are electron-rich, while they are electron poor on acidic supports. Density functional theory calculations for supported platinum particles furthermore revealed that for higher electron richness of the support oxygen atoms the complete Pt density of states shifts to higher energy, that is, lower binding energy, while the location of the interstitial bonding orbital moves from the metal–support interface to the surface of the Pt particle, and that the insulator-to-metal transition, which occurs with increasing Pt nanoparticle size, shifts to lower particle size on basic supports (Oudenhuijzen et al., 2004). The latter implies that the proposed field effect should be more proficient for smaller metal particles, where it also has to be taken into account that both the crystal habit (i.e., which crystal faces are present on the metal nanoparticle; this may be different on a different support; Somorjai et al., 2006) and the possible presence of oxygen (surface or subsurface; Schalow et al., 2006) on the particle may play a role.

Recent work demonstrates the feasibility of a somewhat similar concept as just described, based on a *semiconducting* SnO_2 catalyst, fabricated in the shape of a *nanowire*, and configured as FETs (Zhang et al., 2004b, 2005). The electron density in the nanowire was adjusted electrically, and in this way oxidation and reduction reactions at the surface of the SnO_2 wire could be modified by tuning the density of oxygen vacancies in the surface layer. This work is a very nice demonstration of the merger of different fields, that is, catalysis, nanotechnology, and microelectronics, and it would be quite feasible to combine this with microfluidics. It is quite possible that this is the direction in which the field of microreaction technology in future will develop.

6. CONCLUSION

In this chapter the use of electric fields as a means to control, activate, or modify chemistry in microreactor systems, and in some cases also in downstream work-up microsystems, was discussed. Particular attention was paid to microplasmas, in its many different configurations, because this is an upcoming field which allows us to perform chemistry at low temperatures, which normally can only be performed at high temperatures, with possible advantages in energy efficiency. Novel configurations for plasma generation, based on nanostructured electrodes, were also highlighted. Special attention was also given to electrokinetic control of chemical reactions, a concept which only works efficiently in microsystems, because of the required localization of electric fields, and in some cases also the desired local high density of electric field lines. Electrokinetic concepts can be used to enhance or decrease fluid flow, or to transport and position or trap particles and molecules. A number of less known concepts (electric wind, electric swing adsorption, electrospray, and pulsed electric fields) were discussed, for which it remains to be seen if they will have relevance for microreaction technology. In the final section of this chapter some interesting principles to control the activity and orientation of adsorbed molecules, as well as the adsorption and desorption itself, were described. We predict that these principles will become very relevant in future microreaction technology, where the continuous progress in microsystems integration and in nanotechnology will allow us to combine new activation concepts with the beneficial high surface-to-volume ratio of microstructured systems to develop novel reactor and reaction concepts.

ACKNOWLEDGMENTS

The authors like to express their gratitude to the Technology Foundation STW, applied science division of NWO, and the technology program of the Ministry of Economic Affairs, The Netherlands, for financially supporting the project "Exciting chemistry in microreactors" (VICI "Vernieuwingsimpuls" program, project number 06626). J.G.E. Gardeniers also likes to thank the MESA + Research Institute for Nanotechnology for financial support of the SRO program "Mesofluidics."

REFERENCES

Ağıral, A., Eral, H.B., Mugele, F., and Gardeniers, J.G.E. unpublished results (2010).
Ağıral, A., and Gardeniers, J. G.E. *J. Phys. Chem. C* **112**, 15183 (2008a).
Ağıral, A., Groenland, A. W., Chinthaginjala, J. K., Seshan, K., Lefferts, L., and Gardeniers, J. G.E. *J. Phys. D: Appl. Phys.* **41**, 194009 (2008b).
Ağıral, A., Lefferts, L., and Gardeniers, J. G.E. *IEEE Trans. Plasma Sci.* **37**, 985 (2009).
Ağıral, A., Trionfetti, C., Lefferts, L., Seshan, K., and Gardeniers, J. G.E. *Chem. Eng. Technol.* **31**, 1116 (2008c).
Anderson, T. S., Ma, J. H., Park, S.-J., and Eden, J. G. *IEEE Trans. Plasma Sci.* **36**, 1250 (2008).
Astorga-Wells, J., Bergman, T., and Jörnvall, H. *Anal. Chem.* **76**, 2425 (2004).
Astorga-Wells, J., Vollmer, S., Tryggvason, S., Bergman, T., and Jörnvall, H. *Anal. Chem.* **77**, 7131 (2005).
Baba, K., Okada, T., Kaneko, T., and Hatakeyama, R. *Jpn. J. Appl. Phys.* **45**, 8286 (2006).
Baba, K., Okada, T., Kaneko, T., Hatakeyama, R., and Yoshiki, H. *Thin Solid Films* **515**, 4308 (2007).

Bak, E., Donten, M., Stojek, Z., and Scholz, F. *Electrochem. Commun.* **9**, 386 (2007).

Balint, I., and Aika, K.-I. *J. Chem. Soc., Faraday Trans.* **93**, 1797 (1997).

Banks, C. E., Davies, T. J., Evans, R. G., Hignett, G., Wain, A. J., Lawrence, N. S., Wadhawan, J. D., Marken, F., and Compton, R. G. *Phys. Chem. Chem. Phys.* **5**, 4093 (2003).

Bao, J., and Regnier, F. *J. Chromatogr.* **608**, 217 (1992).

Bazylak, A., Sinton, D., and Djilali, N. *J. Power Sources* **143**, 57 (2005).

Becker, K., Koutsospyros, A., Yin, S.-M., Christodoulatos, C., Abramzon, N., Joaquin, J. C., and Marino, G. B. *Plasma Phys. Control. Fusion* **47**, B513 (2005).

Becker, K. H., Kogelschatz, U., Schoenbach, K. H., Barker, R. (Eds.), Applications of atmospheric-pressure air plasmas, Chapter 9 *in* "Non Equilibrium Air Plasmas at Atmospheric Pressure". IOP Publ., Bristol, UK, 2004.

Becker, K. H., Schoenbach, K. H., and Eden, J. G. *J. Phys. D: Appl. Phys.* **39**, R55 (2006).

Belmont, C., and Girault, H. H. *Electrochim. Acta* **40**, 2505 (1995).

Beltrá, A. P., Bonete, P., González-García, J., García-García, V., and Montiel, V. *J. Electrochem. Soc.* **152**, D65 (2005).

Berge, B. *Comptes Rendus Acad. Sci., Ser. II* **317**, 157 (1993).

Blatter, F., Moreau, F., and Frei, H. *J. Phys. Chem.* **98**, 13403 (1994).

Block, J. *Z. Phys. Chem.* **39**, 169 (1963).

Bose, A. C., Shimizu, Y., Mariotti, D., Sasaki, T., Terashima, K., and Koshizaki, N. *Nanotechnology* **17**, 5976 (2006).

Bragg, A. E., Verlet, J. R.R., Kammrath, A., Cheshnovsky, O., and Neumark, D. M. *Science* **306**, 669 (2004).

Brown, B. D., Smith, C. G., and Rennie, A. R., *Phys. Rev. E* **63**, 016305 (2000).

Brusatori, M. A., Tie, Y., and Van Tassel, P. R., *Langmuir* **19**, 5089 (2003).

Bula, W.P.Microfluidic Devices for Kinetic Studies of Chemical Reactions, Ph.D. thesis, University of Twente (2009).

Burke, B. J., and Regnier, F. *Electrophoresis* **22**, 3744 (2001).

Chang, C.-C., and Yang, R.-J. *Phys. Fluids* **21**, 052004 (2009).

Chatterjee, A. N., and Aluru, N. R. *J. Microelectromech. Syst.* **14**, 81 (2005).

CHEMKIN. "Release 4.0 Software Package". Reaction Design, San Diego, CA (2004).

Chiang, W.-H., and Sankaran, R. M. *Appl. Phys. Lett.* **91**, 121503 (2007).

Chiang, W.-H., and Sankaran, R. M. *J. Phys. Chem. C* **112**, 17920 (2008).

Choi, J. O., Akinwande, A. I., and Smith, H. I. *J. Vac. Sci. Technol. B* **19**, 900 (2001).

Choudhary, T. V., Aksoylu, E., and Goodman, D. W. *Catal. Rev.* **45**, 151 (2003).

Cohen, A. E. *Phys. Rev. Lett.* **94**, 118102 (2005).

Cohen, A. E., and Moerner, W. E. *Appl. Phys. Lett.* **86**, 093109 (2005).

Cohen, A. E., and Moerner, W. E. *Proc. Natl. Acad. Sci.* **103**, 4362 (2006).

Collins, P. G., Arnold, M. S., and Avouris, Ph. *Science* **292**, 706 (2001).

Culbertson, C. T., and Jorgenson, J. W. *Anal. Chem.* **66**, 955 (1994).

Culbertson, C. T., and Jorgenson, J. W. *J. Microcolumn Sep.* **11**, 167 (1999).

Cummings, E. B., Griffiths, S. K., Nilson, R. H., and Paul, P. H. *Anal. Chem.* **72**, 2526 (2000).

De Jong, K. P., and Geus, J. W. *Catal. Rev. Sci. Eng.* **42**, 481 (2000).

De Malsche, W., Eghbali, H., Clicq, D., Vangelooven, J., Gardeniers, H., and Desmet, G. *Anal. Chem.* **79**, 5915 (2007).

De Malsche, W., Op De Beeck, J., Gardeniers, H., and Desmet, G. *J. Chromatogr. A* **1216**, 551 (2009).

Eijkel, J. C.T., Stoeri, H., and Manz, A. *Anal. Chem.* **71**, 2600 (1999).

Ertl, G., Knötzinger, H., and Weitcamp, J. "Handbook of Catalysis". Wiley VCH, Weinheim (1997).

Fletcher, P. D.I., Haswell, S. J., and Paunov, V. P. *Analyst* **124**, 1273 (1999).

Fletcher, P. D.I., Haswell, S. J., Pombo-Villar, E., Warrington, B. H., Watts, P., Wong, S. F.Y., and Zhang, X. *Tetrahedron* **58**, 4735 (2002).

Fletcher, P. D.I., Haswell, S. J., and Zhang, X. *Lab Chip* **2**, 115 (2001).
Foest, R., Schmidt, M., and Becker, K. *Int. J. Mass Spectrom.* **248**, 87 (2006).
Fowler, R. H., and Nordheim, L. *Proc. Roy. Soc. London A* **119**, 173 (1928).
Frei, H. *Science* **313**, 309 (2006).
Fridman, A. F. "Plasma Chemistry". Cambridge University Press, Cambridge (2008).
Garimella, S. V., and Liu, D. J. *Enhanced Heat Transfer* **16**, 237 (2009).
Ghanem, M. A., and Marken, F. *Electrochem. Commun.* **7**, 1333 (2005).
Go, D. B., Fisher, T. S., Garimella, S. V., and Bahadur, V. *Plasma Sources Sci. Technol.* **18**, 035004 (2009).
Go, D. B., Garimella, S. V., Fisher, T. S., and Mongia, R. K. *J. Appl. Phys.* **102**, 053302 (2007).
Grande, C. A., Ribeiro, R. P.P.L., and Rodrigues, A. E. *Energy Fuels* **23**, 2797 (2009).
Green, N. G., Ramos, A., González, A., Morgan, H., and Castellanos, A. *Phys. Rev. E* **61**, 4011 (2000).
Green, N. G., Ramos, A., González, A., Morgan, H., and Castellanos, A. *Phys. Rev. E* **66**, 026305 (2002).
Grotrian, W. *Ann. Physik* **47**, 141 (1915).
Grymonpré, D. R., Finney, W. C., Clark, R. J., and Locke, B. R. *Ind. Eng. Chem. Res.* **43**, 1975 (2004).
Gulder, O. L. *Energy Fuels* **23**, 591 (2009).
Günther, A., Khan, S. A., Thalmann, M., Trachsel, F., and Jensen, K. F. *Lab Chip* **4**, 278 (2004).
Hagelaar, G. J.M., and Pitchford, L. C. *Plasma Sources Sci. Technol.* **14**, 722 (2005).
Halpern, B., and Gomer, R. *J. Chem. Phys.* **51**, 1031 (1969).
Han, B., Li, Z., Pronkin, S., and Wandlowski, T. *Can. J. Chem.* **82**, 1481 (2004).
Harrison, D. J., Fluri, K, Seiler, K, Fan, Z., Effenhauser, C. S., Manz, A. *Science* **261**, 895 (1993).
Hartman, R. L., and Jensen, K. F. *Lab Chip* **9**, 2495 (2009).
Hessel, V., Hardt, S., and Löwe, H. "Chemical Microprocess Engineering. Fundamentals, Modelling and Reactions". Wiley VCH Verlag GmbH & Co. KGaA, Weinheim (2004).
Hessel, V., Renken, A., Schouten, J. C., and Yoshida, J. (Eds.) "Micro Process Engineering. A Comprehensive Handbook (3 volumes)". Wiley VCH Verlag GmbH & Co. KGaA, Weinheim (2009).
Horii, D., Amemiya, F., Fuchigami, T., and Atobe, M. *Chem. Eur. J.* **14**, 10382–10387 (2008).
Hsu, D. D., and Graves, D. B. *Plasma Chem. Plasma Proc.* **21**, 1 (2005).
Huber, D. L., Manginell, R. P., Samara, M. A., Kim, B.-I., and Bunker, B. C. *Science* **301**, 352 (2003).
Jähnisch, K., Hessel, V., Löwe, H., and Baerns, M. *Angew. Chem. Int. Ed.* **43**, 406 (2004).
Janev, R. K., and Reiter, D. *Phys. Plasmas* **11**, 780 (2004).
Jaworek, A., Krupa, A., and Czech, T., *J. Electrostat.* **65**, 133 (2007).
Jayashree, R. S., Gancs, L., Choban, E. R., Primak, A., Natarajan, D., Markoski, L. J., and Kenis, P. J.A. *J. Am. Chem. Soc.* **127**, 16758 (2005).
Jenkins, G., Franzke, J., and Manz, A. *Lab Chip* **5**, 711 (2005).
Jensen, K. F. *Chem. Eng. Sci.* **56**, 293 (2001).
Jiang, X. D., Li, R., Qiu, R., Hu, X., Liang, H. *Chem. Eng. J.* **116**, 149 (2006).
Jortner, J., and Kestner, N. R. (Eds.) "Electrons in Fluids". Springer Verlag, Berlin (1973).
Kadowaki, M., Yoshizawa, H., Mori, S., and Suzuki, M. *Thin Solid Films* **506**, 123 (2006).
Kalman, H., and Sher, E. *Appl. Therm. Eng.* **21**, 265 (2001).
Kane, C., and Tzédakis, T. *AIChE J.* **54**, 1365 (2008).
Kaniansky, D., Masár, M., Marák, J., and Bodor, R. *J. Chromatogr. A* **834**, 133 (1999).
Katayama, H., Honma, H., Nakagawara, N., and Yasuoka, K. *IEEE Trans. Plasma Sci.* **37**, 897 (2009).
Katsura, S., Harada, N., Maeda, Y., Komatsu, J., Matsuura, S.-I., Takashima, K., and Mizuno, A. *J. Biosci. Bioeng.* **98**, 293 (2004).

Kim, S.-O., and Eden, J. G. *IEEE Phot. Tech Lett.* **17**, 1543 (2005).

Kjeang, E., Djilali, N., and Sinton D. *J. Power Sources* **186**, 353 (2009).

Kogelschatz, U. *IEEE Trans. Plasma Sci.* **30**, 1400 (2002).

Kogelschatz, U. *Plasma Chem. Plasma Proc.* **23**, 1 (2003).

Kohlheyer, D., Besselink, G. A.J., Lammertink, R. G.H., Schlautmann, S., Unnikrishnan, S., Schasfoort, R. B.M. *Microfluid. Nanofluid.* **1**, 242 (2005).

Kohlrausch, F. *Ann. Phys. Chem. N. F.* **62**, 209 (1897).

Kolb, G., Hessel, V., Cominos, V., Hofmann, C., Löwe, H., Nikolaidis, G., Zapf, R., Ziogas, A., Delsman, E. R., de Croon, M. H.J.M., Schouten, J. C., de la Iglesia, O., Mallada, R., Santamaria, J. *Catal. Today* **120**, 2 (2007).

Kornienko, O., Reilly, P. T.A., Whitten, W. B., and Ramsey, J. M. *Anal. Chem.* **72**, 559 (2000).

Kranich, A., Ly, H. K., Hildebrandt, P., and Murgida, D. H. *J. Am. Chem. Soc.* **130**, 9844 (2008).

Kreuzer, H. J. *Surf. Interface Anal.* **36**, 372 (2004).

Krivenko, A. G., Komarova, N. S., Piven, N. P. *Electrochem. Commun.* **9**, 2364 (2007).

Krupke, R., Hennrich, F., Kappes, M. M., and Löhneysen, H. V. *Nano Lett.* **4**, 1395 (2004).

Krupke, R., Hennrich, F., von Löhneysen, H., and Kappes, M. M. *Science* **301**, 344 (2003).

Küpper, M., Hessel, V., Löwe, H., Stark, W., Kinkel, J., Michel, M., and Schmidt-Traub, H. *Electrochim. Acta* **48**, 2889 (2003).

Lahann, J., Mitragotri, S., Tran, T.-N., Kaido, H., Sundaram, J., Choi, I. S., Hoffer, S., Somorjai, G. A., Langer, R. *Science* **299**, 371 (2003).

Le Gac, S., and van den Berg, A. (Eds.) "Miniaturization and Mass Spectrometry". The Royal Society of Chemistry, Cambridge, UK (2009).

Lehr, W., and Hiller, W. *J. Electrostat.* **30**, 433 (1993).

Lei, L. C., Zhang, Y., Zhang, X. W., Du, Y. X., Dai, Q. Z., Han, S. *Ind. Eng. Chem. Res.* **46**, 5469 (2007).

Leonard, G. L., Mitchner, M., and Self, S. A. *J. Fluid Mech.* **127**, 123 (1983).

Lerner, E. J. *Industr. Phys.* **6**, 16 (2000).

Leroux, F., Campagne, C., Perwuelz, A., Gengembre, L. *J. Colloid Interf. Sci.* **328**, 412 (2008).

Leroux, F., Perwuelz, A., Campagne, C., and Behary, N. *J. Adhesion Sci. Technol.* **20**, 939 (2006).

Li, J.-R., Kuppler, R. J., and Zhou, H.-C. *Chem. Soc. Rev.* **38**, 1477 (2009).

Lieberman, M. A., and Lichtenberg, A. J. "Principles of Plasma Discharges and Materials Processing". John Wiley, New York (1994).

Locke, B., Sato, M., Sunka, P., Hoffmann, M., and Chang, J. *Ind. Eng. Chem. Res.* **45**, 882 (2006).

Luo, C.-P., Heeren, A., Henschel, W., and Kern, D. P. *Microelectr. Eng.* **83**, 1634 (2006).

MacDonald, S. M., Watkins, J. D., Bull, S. D., Davies, I. R., Gu, Y., Yunus, K., Fisher, A. C., Bulman Page, P. C., Chan, Y., Elliott, C., and Marken, F. *J. Phys. Org. Chem.* **22**, 52 (2009).

MacDonald, S. M., Watkins, J. D., Gu, Y., Yunus, K., Fisher, A. C., Shul, G., Opallo, M., and Marken, F. *Electrochem. Commun.* **9**, 2105 (2007).

MacInnes, J. M., Du, X., and Allen, R. W.K. *Trans. I. Chem. E, Part A.* **81**, Part A, 773 (2003).

Magnier, P., Hong, D. P., Leroy-Chesneau, A., Bauchire, J. M., and Hureau, J. *Exp. Fluids* **42**, 815 (2007).

Markarian, N., Yeksel, M., Khusid, B., and Farmer, K. *Appl. Phys. Lett.* **82**, 4839 (2003).

Marken, F., Webster, R. D., Bull, S. D., and Davies, S. G. *J. Electroanal. Chem.* **437**, 209 (1997).

McEwen, J.-S., and Gaspard, P. *J. Chem. Phys.* **125**, 214707 (2006) Magnier, P., Hong, D. P., Leroy-Chesneau, A., Bauchire, J. M., and Hureau, J. *Exp. Fluids* **42**, 815 (2007).

Meek, S. A., Conrad, H., and Meijer, G. *Science* **324**, 1699 (2009).

Melanson, J. E., Baryla, N. E., and Lucy, C. A. *Trans. Anal. Chem.* **20**, 365 (2001).

Michel, M., Schmidt-Traub, H., Ditz, R., Schulte, M., Kinkel, J., Stark, W., Küpper, M., and Vorbrodt, M. *J. Appl. Electrochem.* **33**, 939 (2003).

Mikkers, F. *Anal. Chem.* **71**, 522 (1999).

Millot, Y., Man, P. P., Springuel-Huet, M.-A., and Fraissard, J. *Comptes Rendus Acad. Sci. Ser. IIC Chem.* **4**, 815 (2001).

Min, S., Evrendilek, G. A., and Zhang, H. Q. *IEEE Trans. Plasma Sci.* **35**, 59 (2007).
Modi, A., Korathar, N., Lass, E., Wei, B., and Ajaya, P. M. *Nature* **424**, 171 (2003).
Mojet, B. L., Miller, J. T., Ramaker, D. E., and Koningsberger, D. C. *J. Catal.* **186**, 373 (1999).
Moon, H., Cho, S. K., Garrell, R. L., and Kim, C.-J. *J. Appl. Phys.* **92**, 4080 (2002).
Moon, S.-H., and Shim, J.-W. *J. Colloid Interf. Sci.* **298**, 523 (2006).
Moreau, E. *J. Phys. D: Appl. Phys.* **40**, 605 (2007).
Mori, S., Yamamoto, A., and Suzuki, M. *Plasma Sources Sci. Technol.* **15**, 609 (2006).
Mugele, F. *Soft Matter* **5**, 3377 (2009).
Mugele, F., and Baret, J.-C. *J. Phys. Condens. Matter* **17**, R705 (2005).
Mugele, F., Baret, J.-C., and Steinhauser, D. *Appl. Phys. Lett.* **88**, 204106 (2006).
Murgida, D. H., and Hildebrandt, P. *J. Phys. Chem. B* **106**, 12814 (2002).
Nenu, R. K.T., Yoshida, H., Fukui, K., and Yamamoto, T. *Powder Technol.* **196**, 147 (2009).
Nichols, K. F.P., and Gardeniers, H. J.G.E. *Anal. Chem.* **79**, 8699 (2007).
Nichols, K.P.F., and Gardeniers, H.J.G.E., Unpublished results (2006).
Niedziolka, J., Szot, K., Marken, F., and Opallo, M. *Electroanalysis* **19**, 155 (2007).
Nozaki, T., Hattori, A., and Okazaki, K. *Catal. Tod.* **98**, 607 (2004).
Nozaki, T., Sasaki, K., Ogino, T., Asahi, D., and Okazaki, K. *Nanotechnology* **18**, 235603 (2007a).
Nozaki, T., Sasaki, K., Ogino, T., Asahi, D., and Okazaki, K. *J. Therm. Sci. Technol.* **2**, 192 (2007b).
Oddy, M. H., Santiago, J. G., and Mikkelsen, J. C. *Anal. Chem.* **73**, 5822 (2001).
Oudenhuijzen, M. K., van Bokhoven, J. A., Ramaker, D. E., and Koningsberger, D. C. *J. Phys. Chem. B* **108**, 20247 (2004).
Paddon, C. A., Atobe, M., Fuchigami, T., He, P., Watts, P., Haswell, S. J., Pritchard, G. J., Bull, S. D., and Marken, F. *J. Appl. Electrochem.* **36**, 617 (2006).
Paik, P., Pamula, V. K., and Fair, R. B. *Lab Chip* **3**, 253 (2003).
Park, S.-J., Kim, K. S., and Eden, J. G. *Appl. Phys. Lett.* **86**, 221501 (2005).
Paschen, F. *Ann. Phys.* **37**, 69 (1889).
Peng, H. Q., Alvarez, N. T., Kittrell, C., Hauge, R. H., and Schmidt, H. K. *J. Am. Chem. Soc.* **128**, 8396 (2006).
Pileni, M.-P. *Adv. Colloid Interf. Sci.* **46**, 139 (1993).
Pohl, H. A. "Dielectrophoresis. The Behavior of Neutral Matter in Non-Uniform Electric Fields". Cambridge University Press, Cambridge (1978).
Pollack, M. G., Shenderov, A. D., and Fair, R. B. *Lab Chip* **2**, 96 (2002).
Pompa, P. P., Bramanti, A., Maruccio, G., Cingolani, R., De Rienzo, F., Corni, S., Di Felice, R., and Rinaldi, R. *J. Chem. Phys.* **122**, 181102 (2005).
Pronkin, S., and Wandlowski, T. J. *Electroanal. Chem.* **550–551**, 131 (2003).
Qian, A., and Bau, H. H. *Anal. Chem.* **74**, 3616 (2002).
Qiao, Q, and Aluru, N. R. *J. Micromech. Microeng.* **12**, 625 (2002).
Raizer, Y. "Gas Discharge Physics". Springer Verlag, Heidelberg (1991).
Rayleigh, L. *Phil. Mag.* **14**, 184 (1882).
Rayner, D., Fietkau, N., Streeter, I., Marken, F., Buckley, B. R., Page, P. C.B., del Campo, J., Mas, R., Munoz, F. X., and Compton, R. G. *J. Phys. Chem. C* **111**, 9992 (2007).
Reuss, R. H., and Chalamala, B. R. *J. Vac. Sci. Technol. B* **21**, 1187 (2003).
Revermann, T., Götz, S., Künnemeyer, J., and Karst, U. *Analyst* **133**, 167 (2008).
Robinson, M. *Am. J. Phys.* **30**, 366 (1962).
Rode, S., and Lapicque, F. Microstructured reactors for electrochemical synthesis, Chapter 17 *in* "Micro Process Engineering. A Comprehensive Handbook. Volume 1, Fundamentals, Operations and Catalysis". Wiley VCH Verlag GmbH & Co. KGaA, Weinheim (2009).
Roth, C. M., Sader, J. E., and Lenhoff, A. M. *J. Colloid Interface Sci.* **203**, 219 (1998).
Ruthven, D. M., and Ching, C. B. *Chem. Eng. Sci.* **44**, 1011 (1989).
Salimi-Moosavi, H., Tang, T., and Harrison, D. J. *J. Am. Chem. Soc.* **119**, 8716 (1997).
Sankaran, R. M., Holunga, D., Flagan, R. C., and Giapis, K. P. *Nano Lett.* **5**, 537 (2005).

Schaffert, R. M. "Electrophotography". John Wiley and Sons, New York (1975).

Schalow, T., Brandt, B., Starr, D. E., Laurin, M., Shaikhutdinov, S. K., Schauermann, S., Libuda, J., and Freund, H.-J. *Angew. Chem. Int. Ed.* **45**, 3693 (2006).

Schasfoort, R. B.M., Schlautmann, S., Hendrikse, L., and van den Berg, A. *Science* **286**, 942 (1999).

Schoenbach, K. H., Moselhy, M., Shi, W., and Bentley, R. *J. Vac. Sci. Technol. A* **21**, 1260 (2003).

Schoenbach, K. H., Verhappen, R., Tessnow, T., Peterkin, P. F., and Byszewski, W. *Appl. Phys. Lett.* **68**, 13 (1996).

Scholz, F., Schröder, U., and Gulaboski, R. "Electrochemistry of Immobilized Particles and Droplets". Springer, Berlin (2005).

Schultz, Z. D., and Gewirth, A. A. *Anal. Chem.* **77**, 7373 (2005).

Sekiguchi, H., Ando, M., and Kojima, H. *J. Phys. D: Appl. Phys.* **38**, 1722 (2005).

Seto, T., Kwon, S.-B., Hirasawa, M., and Yabe, A. *Jpn. J. Appl. Phys.* **44**, 5206 (2005).

Shaik, S., de Visser, S. P., and Kumar, D. *J. Am. Chem. Soc.* **126**, 11746 (2004).

Sheldon, R. A. *Adv. Synth. Catal.* **349**, 1289 (2007).

Sheng, Y.-J., Tsao, H.-K., Zhou, J., and Jiang, S. *Phys. Rev. E* **66**, 011911 (2002).

Shimizu, Y., Sasaki, T., Ito, T., Terashima, K., and Koshizaki, N. *J. Phys. D: Appl. Phys.* **36**, 2940 (2003).

Shimizu, Y., Sasaki, T., Liang, C., Bose, A. C., Ito, T., Terashima, K., and Koshizaki, N. *Chem. Vap. Dep.* **11**, 244 (2005).

Shin, D. H., Kim, J.-E., Shim, H. C., Song, J.-W., Yoon, J.-H., Kim, J., Jeong, S., Kang, J., Baik, S., and Han, C.-S. *Nano Lett.* **8**, 4380 (2008).

Shiu, J.-Y, Kuo, C.-W., and Chen, P. *J. Am. Chem. Soc.* **126**, 8096 (2004).

Shrimpton, J. S., and Crane, R. I. *Chem. Eng. Technol.* **24**, 9 (2001).

Sichler, P., Buttgenbach, S., Hibbe, L. B., Shrader, C., and Gericke, K.-H. *Chem. Eng. J.* **101**, 465 (2004).

Skelton, V., Greenway, G. M., Haswell, S. J., Styring, P., Morgan, D. O., Warrington, B., and Wong, S. Y.F. *Analyst* **126**, 7 (2001a).

Skelton, V., Greenway, G. M., Haswell, S. J., Styring, P., Morgan, D. O., Warrington, B., and Wong, S. Y.F. *Analyst* **126**, 11 (2001b).

Solomentsev, Y., Böhmer, M., and Anderson, J. L. *Langmuir* **13**, 6058 (1997).

Somorjai, G. A., Contreras, A. M., Montano, M., and Rioux, R. M. *Proc. Natl. Acad. Sci.* **103**, 10577 (2006).

Sparreboom, W., van den Berg, A., Eijkel, J. C.T. *Nat. Nanotechnol.* **4**, 713 (2009).

Spindt, C. A., Brodie, I., Humphrey, L., and Westerberg, E. R. *J. Appl. Phys.* **47**, 5248 (1976).

Staack, D., Farouk, B., Gutsol, A., and Fridman, A. *Plasma Sources Sci. Technol.* **14**, 700 (2005).

Stankiewicz, A. I., and Moulijn, J. A. *Chem. Eng. Progr.* **96**, 22 (2000).

Stark, R. H., and Schoenbach, K. H. *J. Appl. Phys.* **85**, 2075 (1999).

Stoukides, M., and Vayenas, C. G. *J. Catal.* **70**, 137 (1981).

Studer, V., Pépin, A., Chen, Y., and Ajdari, A. *Microelectron. Eng.* **61–62**, 915 (2002).

Subramani, H. J., and Kurup, A. S. *Chem. Eng. J.* **120**, 169 (2006).

Suehiro, J., Hidaka, S.-I., Yamane, S., and Imasaka, K. *Sens. Act. B* **127**, 505 (2007).

Suga, S., Okajima, M., Fujiwara, K., and Yoshida, J.-I. *J. Am. Chem. Soc.* **123**, 7941 (2001).

Suzuki, M., Yasukawa, T., Mase, Y., Oyamatsu, D., Shiku, H., and Matsue, T. *Langmuir* **20**, 11005 (2004).

Tachibana, K. *IEEJ Trans. Electr. Electron. Eng.* **1**, 145 (2006).

Tang, C. S., Dusseiller, M., Makohliso, S., Heuschkel, M., Sharma, S., Keller, B., and Vörös, J. *Anal. Chem.* **78**, 711 (2006).

Taniguchi, T., Torii, T., and Higuchi, T. *Lab Chip* **2**, 19 (2002).

Tao, R., Huang, K., Tang, H., and Bell, D., *Energy Fuels* **22**, 3785 (2008).

Tao, R., Huang, K., Tang, H., and Bell, D. *Energy Fuels* **23**, 3339 (2009).

Tas, A., Plasma Induced Catalysis, Ph.D. dissertation, Technische Universiteit Eindhoven, The Netherlands (1995).

Teh, S. Y., Lin, R., Hung, L. H., Lee, A. P. *Lab Chip* **8**, 198 and 3604 (2008).

Thome, B., and Ivory, C. F. *J. Chromatogr. A* **953**, 263 (2002).

Thome, B., and Ivory, C. F. *J. Chromatogr. A* **1129**, 119 (2006).

Thome, B., and Ivory, C. F. *J. Chromatogr. A* **1138**, 291 (2007).

Townsend, J. S. *J. Franklin Inst.* **200**, 563 (1925).

Trau, M., Saville, D. A., Aksay, I. A. *Science* **272**, 706 (1996).

Trionfetti, C., Ağıral, A., Gardeniers, J. G.E., Lefferts, L., and Seshan, K. *ChemPhysChem* **9**, 533 (2008a).

Trionfetti, C., Ağıral, A., Gardeniers, J. G.E., Lefferts, L., and Seshan, K. *J. Phys. Chem. C* **112**, 4267 (2008b).

Trionfetti, C., Babich, I. V., Seshan, K., and Lefferts, L. *Appl. Catal. A* **310**, 105 (2006).

van der Wouden, E. J., Heuser, T., Hermes, D. C., Oosterbroek, R. E., Gardeniers, J. G.E., and van den Berg, A. *Colloids Surf. A* **267**, 110 (2005).

Van Krieken, M., and Buess-Herman, C. *Electrochim. Acta* **45**, 675 (1999).

Vayenas, C. G., Bebelis, S., and Ladas, S. *Nature* **343**, 625 (1990).

Vayenas, C. G., and Koutsodontis, C. *J. Chem. Phys.* **128**, 182506 (2008).

von Allmen, P., McCain, S. T., Ostrom, N. P., Vojak, B. A., Eden, J. G., Zenhausern, F., Jensen, C., and Oliver, M. *Appl. Phys Lett.* **82**, 2562 (2003).

von Engel, A. "Ionized Gases". Clarendon Press, Oxford (1955).

Wang, H., Castner, D. G., Ratner, B. D., and Jiang, S. *Langmuir* **20**, 1877 (2004).

Wang, J.-X., and Lunsford, J. H. *J. Phys. Chem.* **90**, 5883 (1986).

Wang, X., Zhou, D., Sinniah, K., Clarke, C., Birch, L., Li, H., Rayment, T., and Abell, C. *Langmuir* **22**, 887 (2006).

Watts, P., Haswell, S. J., Pombo-Villar, E. *Chem. Eng. J.* **101**, 237 (2004).

Wheeler, A. R., Moon, H., Kim, C.-J., Loo, J. A., and Garrell, R. L. *Anal. Chem.* **76**, 4833 (2004).

Wong, P. K., Chen, C.-Y., Wang, T.-H., and Ho, C.-M. *Anal. Chem.* **76**, 6908 (2004).

Yamatake, A., Fletcher, J., Yasuoka, K., and Ishii, S. *IEEE Trans. Plasma Sci.* **34**, 1375 (2006).

Yan, K., Li, R., Zhu, T., Zhang, H., Hu, X., Jiang, X., Liang, H., Qiu, R., and Wang, Y. *Chem. Eng. J.* **116**, 139 (2006).

Yang, Q., and Zhong, C. *ChemPhysChem* **7**, 1417 (2006).

Yoon, J.-Y., and Garrell, R. L. *Anal. Chem.* **75**, 5097 (2003).

Yoon, S. K., Choban, E. R., Kane, C., Tzedakis, T., and Kenis, P. J.A. *J. Am. Chem. Soc.* **127**, 10466 (2005).

Yoon, S. K., Fichtl, G. W., and Kenis, P. J.A. *Lab Chip* **6**, 1516 (2006).

Yoshida, J.-I. "Flash Chemistry. Fast Organic Synthesis in Microsystems". John Wiley & Sons, Ltd., Chichester (2008).

Yoshida, J.-I., Kataoka, K., Horcajada, R., and Nagaki, A. *Chem. Rev.* **108**, 2265 (2008).

Yoshitake, H., and Iwasawa, Y. *J. Phys. Chem.* **96**, 1329 (1992).

Zeng, X. A., Yu, S. J., Zhang, L., and Chen, X. D. *Innov. Food Sci. Emerging Technol.* **9**, 463 (2008).

Zhang, W., Fisher, T. S., and Garimella, S. V. *J. Appl. Phys.* **96**, 6066 (2004a).

Zhang, Y., Kolmakov, A., Chretien, S., Metiu, H., and Moskovits, M. *Nano Lett.* **4**, 403 (2004b).

Zhang, Y., Kolmakov, A., Lilach, Y., and Moskovits, M. *J. Phys. Chem. B.* **109**, 1923 (2005).

Zhao, Y., and Cho, S. K. *Lab Chip* **7**, 273 (2007).

CHAPTER **3**

High-Throughput Organic Synthesis in Microreactors

Charlotte Wiles[1,2,*] and **Paul Watts**[1,*]

Contents			

Abstract   With an average lead time of 10–12 years for a compound to progress from initial identification through clinical trials and finally into a medicine, the pharmaceutical industry are interested in the development of techniques which have the potential to reduce the time taken to generate prospective lead compounds and translate the protocols into production. As such, one of the areas of synthetic chemistry that has benefited greatly from microreaction technology (MRT), over the past 15 years, has been that of pharmaceutical

1   Department of Chemistry, The University of Hull, Cottingham Road, Hull HU6 7RX, UK

2   Chemtrix BV, Burgemeester Lemmensstraat 358, 6163JT Geleen, The Netherlands

* Corresponding author.
E-mail address: c.wiles@chemtrix.com
E-mail address: p.watts@hull.ac.uk

Advances in Chemical Engineering, Volume 38
ISSN: 0065-2377, DOI 10.1016/S0065-2377(10)38003-3

© 2010 Elsevier Inc.
All rights reserved.

research and development (Glasnov and Kappe, 2007; Mason et al., 2007); with interest in the technology stemming from the perceived ease with which reaction conditions can be optimized and subsequently employed across a range of substrates in order to generate compound libraries (Kirschning et al., 2006; Wiles and Watts, 2007a; Wiles and Watts 2008a). The rapid translation of reaction methodology from the microreactors employed within an R&D facility to production, achieved by a process referred to as scale-out, numbering-up, or upscaling, also has the potential to reduce the time taken to take a compound to market (Yoshida, 2008).

Based on this notion, research into the use of microreactors as tools for high-throughput organic synthesis has grown in popularity, with the last 5 years seeing an emerging trend in the types of compounds prepared changing from proof of concept reactions, to the synthesis of molecules of direct relevance to the pharmaceutical industry. With this in mind, the chapter begins with an overview of the seminal examples that helped to foster interest in microreactors as high-throughput tools, illustrating liquid phase (Sections 2.1–2.3), catalytic (Section 2.4) and photochemical (Section 2.6) reactions, and concludes with a selection of current examples into the synthesis of industrially relevant molecules using MRT (Section 3).

1. INTRODUCTION

When considering the use of microreaction technology, from the perspective of high-throughput organic synthesis, the main benefit that this technique offers is increased reaction control, which in itself affords many practical advantages to the user. As a result of the small reactor dimensions, rapid mixing of reactants, and an even temperature distribution are observed, which not only increase the uniformity of reaction conditions, but also afford increased reaction safety, selectivity, reproducibility, and efficiency when compared to conventional batch reactors; where hot spot formation can lead to the formation of by-products and the risk of thermal runaway.

From a screening perspective, the rapid mixing and equilibration of reaction conditions is advantageous as parameters such as time, temperature, pressure, and stoichiometry can be varied with ease and have an immediate effect on the microreaction, compared to batch vessels where changes take time to have an effect on the whole of the reaction mixture. As such, reaction screening can be performed in a shorter time frame, using less raw materials than a conventional reaction evaluation; whilst enabling the reaction conditions identified to be employed in the production of larger quantities of the material for clinical trials and subsequent production campaigns if required. With these factors in mind, the following sections provide practical examples of the advantages associated with microreaction technology (MRT), starting with liquid-phase reactions (Sections 2.1–2.3) and moving on to those reactions that have employed solid-supported reagents, catalysts, and scavengers (Section 2.4–2.5) into the continuous flow process.

2. THE USE OF MICROREACTORS FOR HIGH-THROUGHPUT ORGANIC SYNTHESIS

2.1 Enhanced reaction control

An early example that illustrated the ability to reduce reaction times as a consequence of conducting reactions in miniaturized, continuous flow reactors was reported by Fernandez-Suarez et al. (2002) who investigated a series of Domino reactions within a soda-lime glass microreactor. As Scheme 1 illustrates, the reactions involved a base-catalyzed Knoevenagel condensation followed by an intramolecular hetero-Diels–Alder reaction. To synthesize cycloadduct **1** under continuous flow, a solution of *rac*-citronellal **2** (0.10 M) and a premixed solution containing barbituric acid **3** (0.12 M) and ethylenediamine diacetate (EDDA) **4** (0.17 M), in MeOH:H$_2$O (80:20), were introduced into the reaction channels from inlets A and B, respectively. Reagents were mobilized within the reactor using a series of computer-controlled pumps, with reactions performed over a period of 30 min prior to analysis of the reaction products off-line by liquid chromatography–mass spectrometry (LC–MS). Employing an initial residence time of 2 min, the authors reported 60% conversion to cycloadduct **1**, which was further increased to 68% conversion by extending the residence time to 6 min. With this information in hand, the authors synthesized a series of cycloadducts under the aforementioned conditions, affording the respective compounds in yields ranging from 59 to 75%.

In another glass (borosilicate) microreactor [channel dimensions = 350 μm (wide) × 52 μm (deep) × 2.5 cm (long)], Wiles et al. (2004b) prepared a series of 1,2-azoles, illustrating the synthesis of a pharmaceutically relevant core motif. Reactions were performed using electroosmotic flow (EOF) as the pumping mechanism and employed separate

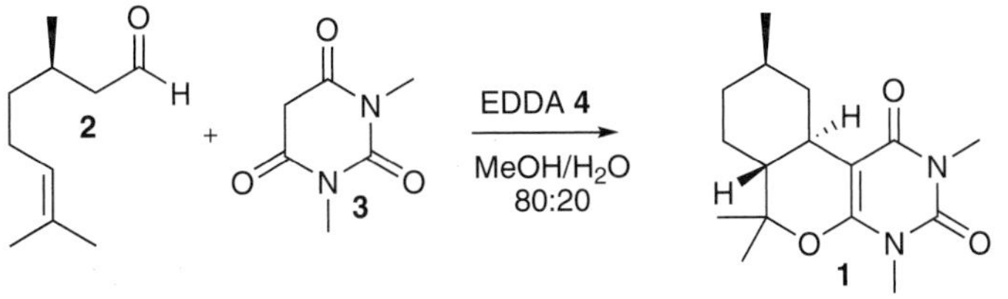

Scheme 1 An example of the Domino reactions performed by Fernandez-Suarez et al. (2002) in a soda-lime microreactor.

reactant solutions of 1,3-diketones (1.0 M) and the hydrazine derivative (1.0 M) in anhydrous THF.

Application of a positive voltage resulted in pumping of the reactants through the microchannel network, into a central channel where the reaction occurred and toward the reactor outlet ($0 \, V \, cm^{-1}$). Employing a range of applied fields enabled the reactant residence time to be optimized, resulting in the synthesis of numerous 1,2-azoles in excellent purity, compared to analogous batch reactions. As Table 1 illustrates, reactions involving hydrazine monohydrate ($R^3 = H$) afforded excellent conversions; however, when a substituted hydrazine derivative, benzyl hydrazine hydrochloride ($R^3 = CH_2Ph$), was employed only 42% conversion to the target 1,2-azole was obtained. This was later optimized to 100% by the use of a stopped flow technique, which served to increase the reactant residence time within the microreactor. Should the technique be required for the production of larger quantities of material, the same effect could be obtained by increasing the length of the microchannel in order to obtain an increased residence time under continuous flow.

Employing a staked plate microreactor (channel dimensions $= 100 \, \mu m$, total volume $= 2 \, ml$), Acke and Stevens (2007) reported the continuous flow synthesis of a series of chromenones via a multicomponent route consisting of a sequential Strecker reaction-intramolecular nucleophilic addition and tautomerization, as depicted in Scheme 2.

To synthesize these pharmaceutically interesting compounds, methanolic solutions of acetic acid 6 (2.0 eq.)/2-formylbenzoic acid 7 (1.0 eq.) and aniline 8 (2.0 eq.)/potassium cyanide 9 (1.2 eq.) were introduced into the reactor from separate inlets, thus ensuring the formation of HCN and the

Table 1 A summary of the results obtained for the synthesis of 1,2-azoles using EOF as the pumping mechanism

R	R^1	R^2	R^3	Applied field (V cm^{-1})[a]	Conversion (%)
CH_3	H	CH_3	H	292, 318	100
Ph	H	CH_3	H	364, 341	100
$-(CH_2)_4-$		Ph	H	260, 303	100
Ph	H	Ph	H	386, 364	100
Ph	CH_3	CH_3	H	292, 318	100
Ph	H	CH_3	CH_2Ph	318, 318	42 (100)[b]

[a]All reaction products were collected at $0 \, V \, cm^{-1}$.
[b]Stopped flow was used.

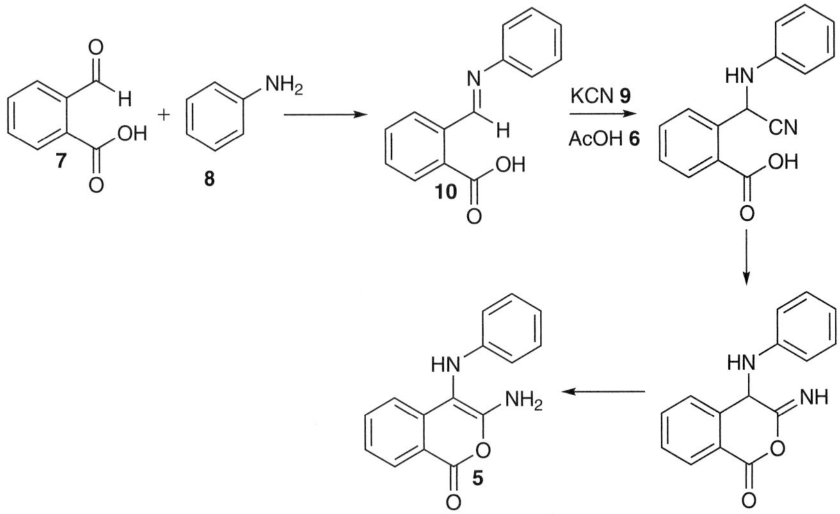

Scheme 2 Reaction protocol employed for the synthesis of 3-diamino-1H-isochromen-1-one **5** under continuous flow conditions.

imine **10** occurred within the confines of the microchannel. In order to prevent precipitation of the reaction products within the residence time unit, and the reactor outlet, a maximum concentration of 0.15 M was selected for the 2-formyl benzoic acid **7**. Employing a residence time of 40 min, the authors reported an isocoumarin **5** yield of 66%, which equates to a throughput of 1.80 g h^{-1}; as the target materials were found to be unstable in solution, the chromenones were stored in a solid form. To demonstrate the flexibility of the reaction setup, a series of amines were employed in place of aniline **8** to afford the target products in yields ranging from 6 to 75% (Table 2). Although these yields may be viewed as moderate to good, several of the examples reported demonstrated an increase in yield compared to analogous batch reactions.

Building on this research, Acke et al. (2008) extended their investigation to include the ring closing of vicinal amino groups in order to introduce an imidazole core into the compounds skeleton. Employing the synthesis of 1H-isochromeno[3,4-d]imidazol-5-one **11** as a model reaction (Scheme 3), the authors investigated the effect of solvent, temperature, acid/orthoester stoichiometry and reactant concentration within a stacked plate microreactor.

Using this approach, the authors found the optimal reaction conditions to be a reactant concentration of 0.2 M, dimethylformamide (DMF) as the reaction solvent, a reaction temperature of 22 °C, 10 mol% p-TsOH **12**, and 5.0 eq. of the orthoester **13**. Under the aforementioned conditions, the target material **11** was obtained in 43% yield, with a residence time of

Table 2   A selection of chromenones synthesized under flow by Acke et al. (2007)

Amine	Yield (%)	Throughput (g h^{-1})
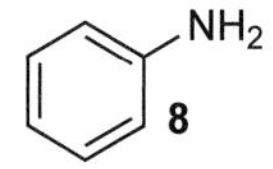 [structure with NH$_2$, labeled **8**]	66	1.80
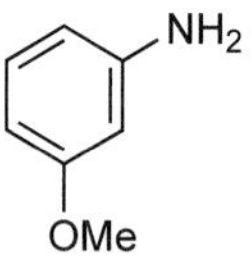 [structure with NH$_2$, OMe]	75	2.28
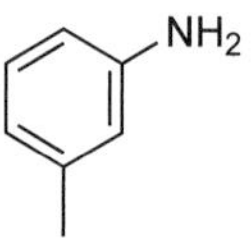 [structure with NH$_2$]	9	1.98
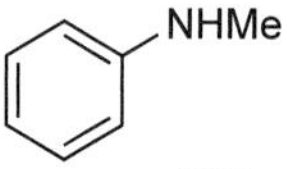 [structure with NHMe]	49	1.41
 [structure with NH$_2$]	6	0.14

Scheme 3   Illustration of the model reaction employed to evaluate the ring closure of 3-amino-4-(arylamino)-1H-isochromen-1-ones under continuous flow.

19.6 min, which was successfully increased to 80% as a result of increasing the residence time to 59 min. Although quantitative conversion to **11** was obtained, the use of DMF as the reaction solvent led to a relatively large loss of product **11** upon isolation. However, having demonstrated the ability to perform the desired ring closure, the authors subsequently evaluated the generality of the protocol for a series of substituted isochromen-1-ones, as illustrated in Table 3, affording the target compounds in a continuous output of 0.18–2.21 g h^{-1}.

The conversion of carboxylic acids to esters is a fundamental transformation in the synthetic chemists' toolbox, as such Wiles et al. (2003) investigated the ability to efficiently synthesize a series of methyl, ethyl,

Table 3 Summary of the 1*H*-isochromeno[3,4-*d*]imidazol-5-ones synthesized in a stacked plate microreactor using *p*-TsOH **12** as the promoter

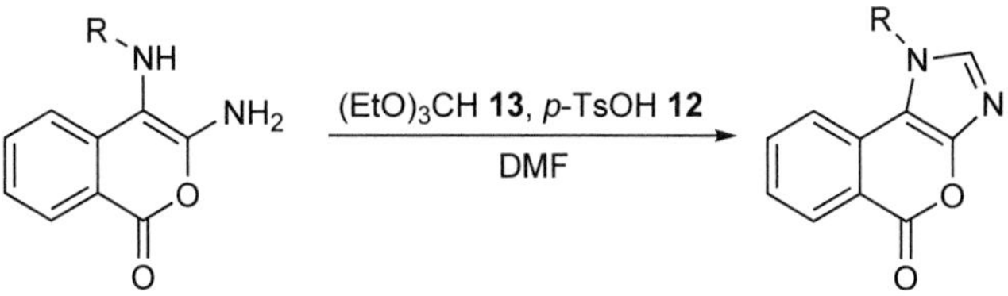

R	Yield (%)[a]	Throughput (g h^{-1})
3-Methoxyphenyl	55	0.39
3-Tolyl	27	0.18
4-Tolyl	78	0.52
4-Methoxyphenyl	75	0.53
4-Fluorophenyl	92	0.62

[a]Residence time = 118 min.

and benzyl ester under continuous flow, using EOF as the pumping mechanism. Employing a borosilicate glass microreactor [channel dimensions = 350 μm (wide) × 52 μm (deep) × 2.5 cm (long)] with three inlets and one outlet, the authors investigated the use of a mixed anhydride, to synthesize esters from carboxylic acids, as depicted in Scheme 4.

Solutions of triethylamine (Et$_3$N) **14** (1.0 M), premixed carboxylic acid/alkyl chloroformate (1.0 M respectively), and 4-dimethylaminopyridine **15** (0.5 M) in MeCN were introduced into the reactor from separate inlets and the reaction products collected at the outlet in MeCN, prior to analysis by gas chromatography–mass spectrometry (GC–MS). Under optimized reaction conditions, the authors were able to synthesize the methyl **16**, ethyl **17**, and benzyl **18** esters in quantitative conversion, with no anhydride or deprotection by-products detected (as observed in conventional batch reactions). In addition to the Boc-glycine derivatives illustrated in Scheme 4, the authors also esterified a series of aromatic carboxylic acids with yields ranging from 91 to 100%, depending on the additional functional groups present.

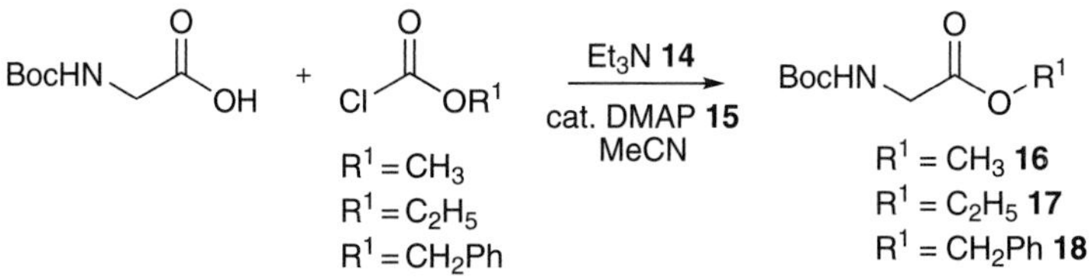

Scheme 4 A selection of the esters synthesized in an EOF-based microreactor.

Of the homogeneous reactions investigated within microstructured reactors, the acylation of amines represents one of the most frequently reported classes of reaction observed to benefit from continuous processing (Kikutani et al. 2002a, 2002b; Schwalbe et al., 2002, 2004). In a recent example by Hooper and Watts (2007), the atom efficiency of reactions conducted within microfluidic systems was demonstrated via the incorporation of deuterium labels into an array of small organic compounds. Employing the base-mediated acylation of primary amines as a model reaction, see Scheme 5, the authors demonstrated the ease by which reactions could firstly be conducted using unlabeled precursors and once the optimal reaction conditions were identified, substitution with the labeled reagent enabled the rapid and efficient synthesis of the deuterated analog.

To conduct the model reaction illustrated, two borosilicate glass reactors were employed in series [Reactor $1 = 201 \, \mu m$ (wide) $\times 75 \, \mu m$ (deep) $\times 2 \, cm$ (long), Reactor $2 = 158 \, \mu m$ (wide) $\times 60 \, \mu m$ (deep) $\times 1.5 \, cm$ (long)] and reagents delivered to the reaction channels using a syringe pump. To ensure long-term operation of the reactor setup, the authors found it necessary to employ a mixed solvent in order to obtain a balance between by-product solubility (Et$_3$N·HCl) and stability of the acylating agent 19. With this in mind, solutions of benzylamine 20 and triethylamine 14 in MeCN (0.1 M, respectively) were introduced into the reactor from separate inlets and mixed using a T-mixer, prior to the addition of the acyl halide (0.05 M) as a solution in anhydrous THF. Employing a total flow rate of $40 \, \mu l \, min^{-1}$, equating to a residence time of 2.6 s, the authors obtained N-benzamide 21 in 95% conversion [determined by off-line high-performance liquid chromatography (HPLC) analysis].

The ease of method transfer was subsequently demonstrated via substitution of acetyl chloride 19 with acetyl [D$_3$]chloride 22, whereby operating the reactor under the aforementioned conditions, the authors obtained comparable results (98% conversion). This investigation provided important results for those in the area of isotope labeling as it demonstrated the ability to perform and optimize reactions using cheap, readily available precursors and then substitute labeled precursors to obtain the respective

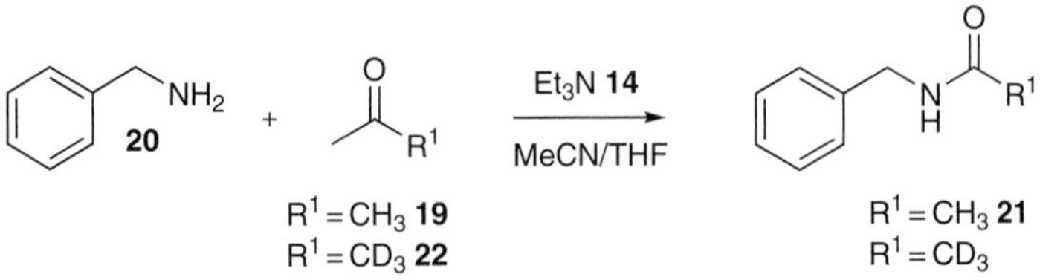

R^1 = CH$_3$ 19
R^1 = CD$_3$ 22

R^1 = CH$_3$ 21
R^1 = CD$_3$

Scheme 5 Schematic illustrating the methodology employed for the incorporation of deuterium labels into small organic compounds.

labeled analog. For applications such as this, microreaction technology affords a means of reducing the development costs associated with the production of small molecule libraries, whilst also affording a rapid and efficient means of producing the labeled compounds.

2.2 Thermal control

Through the examples discussed thus far it can be seen that reaction times can be dramatically reduced as a result of moving from a batch vessel to a microreactor, typically seconds or minutes, a feature that can be attributed to the increased mixing efficiency and excellent thermal control obtained through the efficient dissipation of heat generated during a reaction. This makes microreactors ideal vessels to perform reactions that are temperature sensitive, along with those that are highly exothermic and would prove too hazardous to be performed batchwise. In addition, in the case of those reactions that require activation, thermal control is required in order to heat the microreactors and as the following section illustrates, this has been achieved by the use of thermal jackets, baths, and microwave irradiation.

2.2.1 Increased reaction temperature and pressure

van Meene et al. (2006) demonstrated the ability to conduct reactions above room temperature within a microreactor, reporting the synthesis of α-aminophosphonates via the Kabachnick–Fields reaction (Scheme 6). Using a stainless-steel flow reactor, maintained at 50 °C, the authors found that employing a residence time of 78 min (600 μl min^{-1}) afforded quantitative conversion of the aldimine **23** to the respective α-aminophosphonate **24** and after subjecting the reaction products to an off-line acid–base extraction, an isolated yield of 94% **24** was obtained. Under the aforementioned reaction conditions, a further four derivatives were synthesized, with isolated yields ranging from 68 to 91%.

Compared to conventional batch protocols, where catalysts such as LiClO$_4$ and InCl$_3$ are frequently employed, the continuous flow approach proved advantageous as the reaction was conducted in the absence of a catalyst and without the need for lengthy reaction times, maintained at reflux.

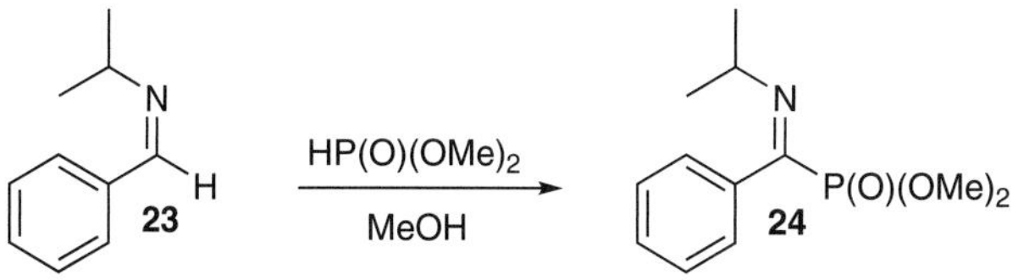

Scheme 6 Example of the α-aminophosphonates synthesized under continuous flow.

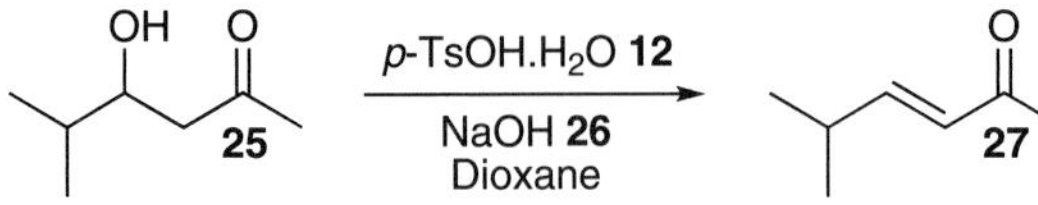

Scheme 7   Acid-catalyzed elimination used to dehydrate β-hydroxyketones.

Tanaka et al. (2007) recently demonstrated the use of an IMM micromixer for an investigation into the development of a continuous flow method for the dehydration of β-hydroxyketones to the target unsaturated products, as depicted in Scheme 7. A typical protocol involved the introduction of 4-hydroxy-5-methylhexan-2-one **25** (1×10^{-2} M) and p-TsOH. H_2O **12** (1×10^{-2} M) in dioxane into the micromixer, the reaction mixture was then heated in a microchannel prior to quenching the resulting reaction mixture with aq. NaOH **26** (1.0 M). Using this approach, the authors found the optimal reaction conditions to be a residence time of 47 s (200 μl min^{-1}) and a reaction temperature of 110 °C, which afforded the target alkene **27** in quantitative yield. In comparison, batch reactions typically afforded a respectable 71% yield; however, the product **27** was accompanied by a large proportion of by-products. Compared to previous batch methodology, the continuous flow approach provides a facile approach for the dehydration of β-hydroxyketones and as discussed in Section 3. (Scheme 71), enables production quantities of synthetically useful material to be generated with ease.

One approach that is becoming more common within research laboratories, and proves impractical when considering the need to scale a process, is the activation of reactions by employing high temperatures and pressure. However through the use of microreaction technology, such reaction conditions can be employed safely and with ease, as the high-temperature/pressure regime recently employed by Hessel et al. (2005) for the Kolbe–Schmitt reaction of resorcinol **28**, to afford 2,4-hydroxybenzoic acid **29** (Scheme 8) demonstrates. As the reaction is industrially relevant, potassium hydrogen carbonate **30** was selected as it is a cheap, readily available raw material making it feasible to use in the large-scale synthesis of **29** (100–1,000 l h^{-1}); furthermore, water was selected as the reaction medium as it is a safe and inexpensive solvent.

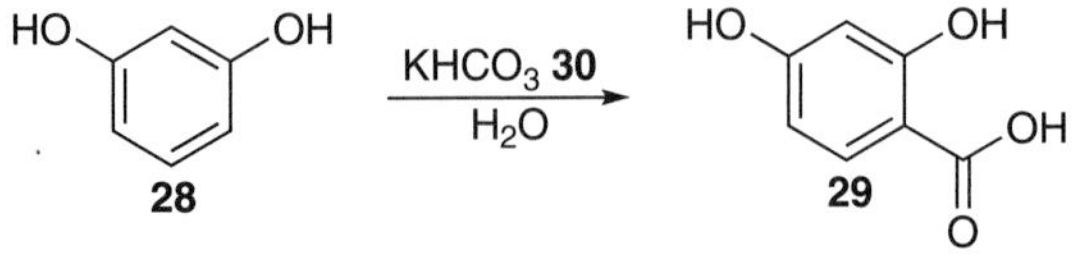

Scheme 8   Kolbe–Schmitt synthesis of 2,4-dihydrobenzoic acid **29** from resorcinol **28**.

Utilizing a stainless-steel capillary-based reactor, the authors initially evaluated the effect of pressure on a reaction conducted at 120 °C and found that with a fixed residence time of 6.5 min an increase in yield from 23 to 47% was observed upon increasing reactor pressure from 1 to 32 bar. In an analogous batch reaction, conducted at reflux, the authors obtained 2,4-dihydroxybenzoic acid **29** in ∼40% yield, after 2 h, with only 90% selectivity. By overpressuring the reaction vessel to 40 bar, the authors were able to increase the boiling point of water to 250 °C, enabling the evaluation of reaction temperature to be conducted over a wide range with relative ease, without encountering a complex biphasic reaction mixture. Maintaining a moderate overpressure, the effect of reaction temperature was subsequently investigated, with results comparable to batch obtained at 140 °C and a residence time of 12 min. The use of higher reaction temperatures was however found to be disadvantageous as it led to decomposition of the target benzoic acid **29** via decarboxylation, the technique did, however, afford a 440-fold increase in the space time yield (cf. a 1 l flask) representing a throughput of 110 g **29** h^{-1}.

Kawanami et al. (2007) demonstrated the ability to perform a series of copper-free Sonogashira C–C coupling reactions in water through the use of a superheated Hastelloy micromixer [0.5 mm (i.d.)] and tubular flow reactor [1.7 mm (i.d.) × 10 m (length)]. Employing a flow process consisting of rapid collision mixing between a substrate and water (to afford particle dispersion) followed by rapid heating (to induce reaction) and cooling to afford a binary phase (consisting of reaction products and water), the authors were able to obtain the target compounds in excellent yield and selectivity when compared to the use of organic solvents. Conducting the reaction illustrated in Scheme 9 at 250 °C and 16 MPa, near quantitative coupling of 4-iodobenzene **31** to phenylacetylene **32** was obtained employing 2 mol% of catalyst **33** in 0.1–4.0 s. Upon collection, the reaction products floated on the surface of the water and the catalyst precipitated as Pd⁰, as such the diphenylacetylene **34** was isolated by a facile process of filtration and phase separation.

In addition to the rate acceleration observed, the technique of rapid mixing and heating afforded an unprecedented catalytic turnover frequency of $4.3 \times 10^6 \, h^{-1}$. Using this approach, the authors subsequently investigated the C–C coupling reactions between phenyl acetylene **32** and a further nine aryl halides to afford yields ranging from 62 to 100%.

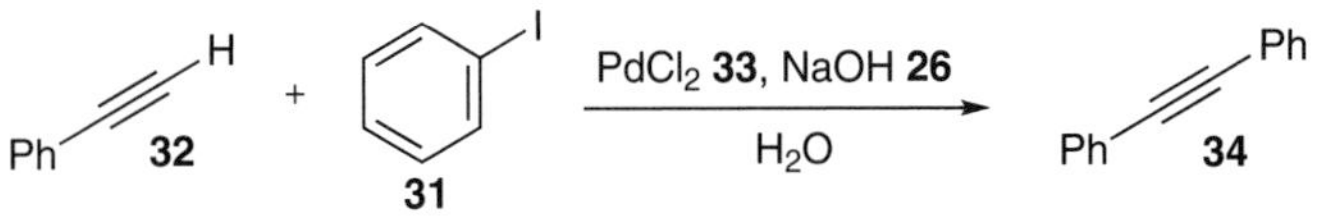

Scheme 9 Model reaction used to demonstrate the rate acceleration attained using superheated water as a reaction solvent.

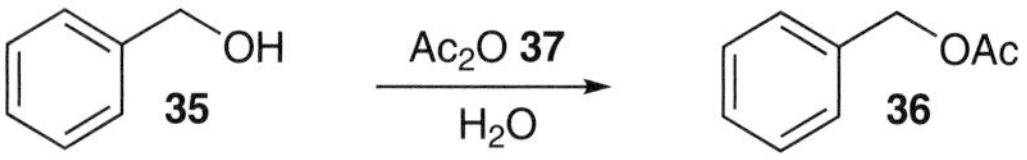

Scheme 10   An example of the hydrolysis free O-acylation conducted in subcritical water.

In a second example, Sato et al. (2007) demonstrated the highly selective, hydrolysis free O-acylation of alcohols using subcritical water (Scheme 10). Employing a reaction temperature of 200 °C, a pressure of 5 MPa, the acylation of benzyl alcohol **35** was achieved in <10 s, using acetic anhydride **37** as the acylating agent, and afforded the target benzyl acetate **36** in 99% yield, compared to 17% in a batch process. The use of subcritical water coupled with the microflow reactor has great synthetic potential as it removes the need for organic solvents enabling reactions to be performed in an environmentally benign solvent system, whilst affording a facile means of isolating the reaction products, facilitating reuse of the solvent.

In addition to providing heat to a reaction vessel through the use of a thermal jacket, oil bath, or convective heater, the past 5 years have seen numerous authors successfully combine the emerging technologies of microwave chemistry and microreaction technology, noting an array of advantages including reduced reaction times and increased product selectivity. Comer and Organ (2005a) recently demonstrated the use of microwave irradiation as an alternative heating source for a series of Suzuki–Miyaura reactions conducted in glass capillary reactors [200 μm (i.d)]. To perform a reaction, the authors introduced reactants into the flow reactor through a stainless-steel mixing chamber, under pressure-driven flow, and investigated the coupling of 4-iodooct-4-ene **38** (0.20 M) and 4-methoxyboronic acid **39** (0.24 M) using THF as the reaction solvent and palladium tetrakis(triphenylphosphine) **40** as the catalyst (Scheme 11). Under 100 W, the authors obtained 100% conversion to 1-methoxy-4-(1-propylpent-1-enyl) benzene **41** with a residence time of 28 min.

The reaction was subsequently repeated using an array of aryl halides and boronic acids, affording the target compounds in 37% to quantitative yield. Compared to conventional batch reactions, these results illustrate dramatic improvements in yield largely due to the suppression of competing side reactions.

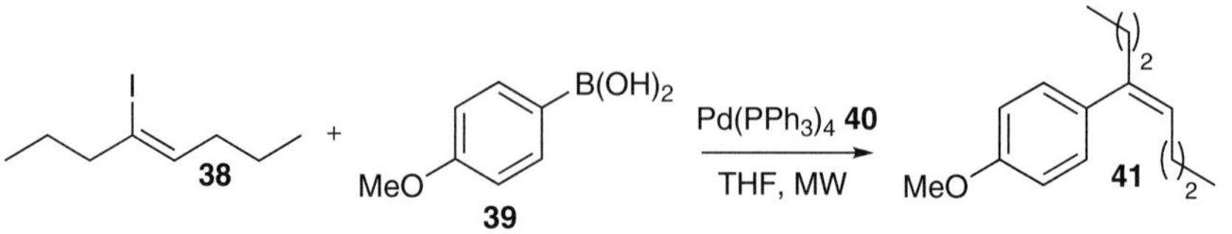

Scheme 11   Schematic illustrating the Suzuki–Miyaura reaction, using microwave heating.

Other reactions evaluated by the authors include the Wittig–Horner olefination and a series of ring-closing metathesis reactions, employing Grubb's II catalyst; in all the cases, a reduction in power consumption, increase in yield and reduction in reaction time was obtained as a result of employing microwave-assisted continuous flow reactions (Comer and Organ, 2005b).

Using a coiled perfluoroalkoxy alkane tube, with an internal diameter of 750 µm (total volume = 3 ml), Benali et al. (2008) investigated the effect of microwaves on a bromination reaction (Scheme 12), the product of which was required in a large amount for a drug discovery program. Employing microwave power of 300 W, affording a reaction temperature of 120 °C, coupled with a flow rate of 0.65 ml min^{-1}, the authors were able to obtain the target compound in 89% yield and 91% purity. Unfortunately, the time taken for the system to reach a steady state, due to the reactor volume, resulted in a large amount of waste generation and the authors sought an alternative method for microwave reaction optimization.

To increase the rate of optimization and reduce the volume of material used, the authors subsequently investigated the formation of discreet sample plugs separated by a fluorous solvent, perfluoromethyldecalin, enabling a large number of reactions to be performed in series, each with a relatively small reactant volume. Initial investigations therefore focused on the halide displacement reaction illustrated in Scheme 13 and importantly identified that analogous conversions were obtained across plug volumes ranging from 200 to 4000 µl (~85% conversion to **42**);

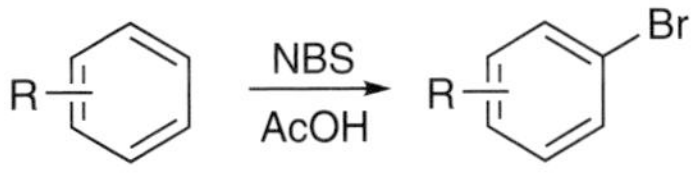

Scheme 12 Bromination reactions used to evaluate continuous flow microwave-assisted reaction optimization.

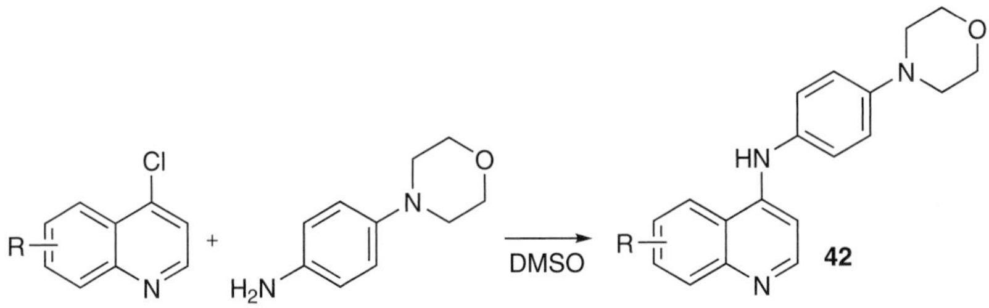

Scheme 13 Microwave-assisted displacement of an aromatic chloride by a 2° amine used to evaluate the effect of plug size in the continuous flow reactor.

meaning that reactions could readily be scaled once optimized using 200 μl volumes of reactants.

The facile approach to reaction optimization, followed by scale-out was subsequently demonstrated using the Suzuki–Miyaura cross-coupling reaction. As Table 4 illustrates, excellent results were obtained using a range of aryl bromides, with comparable or better yields obtained than the analogous batch reactions. Employing a constant reaction stream for a period of 35 min, equivalent to a reaction volume of 30 ml, the authors were able to synthesize a target biphenyl in an overall yield of 58%, corresponding to a throughput of $0.5\,\mathrm{g\,h^{-1}}$.

Table 4 Suzuki–Miyaura reaction used as a model to demonstrate the optimization of microwave induced continuous flow reactions

Aryl bromide	Yield (%)	
	Batch	Flow
	100	100
	100	91
	100	96
	49	100
	100	86

2.2.2 Reduced reaction temperatures

In addition to introducing energy into reactors via heating, or removal of heat generated during a reaction, some synthetic processes need to be conducted at reduced temperatures in order to prevent decomposition of unstable intermediates or to preserve reaction stereoselectivity.

An early example of a microreaction conducted at a reduced reaction temperature was demonstrated by Wiles et al. (2004c) and was used to illustrate the stereoselective alkylation of an Evans auxiliary **43** under continuous flow, as depicted in Scheme 14. Employing a borosilicate glass microreactor [microchannel = 152 μm (wide) × 51 μm (deep) × 2.3 cm (long)] and microtee (Upchurch Scientific, USA), submerged in a solid CO_2/ether ice bath ($-100\,°C$), the authors investigated the deprotonation of 4-methyl-5-phenyl-3-propionyloxazolidin-2-one **43**, using NaHMDS **44** (in anhydrous THF) within a T-reactor. The *in situ* generated enolate **45** was subsequently reacted with benzyl bromide **46** and the reaction products collected at room temperature where immediate quenching was employed. Analysis of the resulting reaction mixture, by GC–MS, enabled the conversion of **43** to diastereomers **47** and **48** to be assessed and the ratio of diastereomers quantified.

Using a total flow rate of $30\,\mu l\,min^{-1}$, the authors obtained 41% conversion to **47** and **48** and a diastereoselectivity of 91:9 (**47:48**), with 59% residual N-acyl oxazolidinone **43**. In an analogous batch reaction, conducted at $-100\,°C$, the diastereomers **47** and **48** were obtained in an overall yield of 68% and a ratio of 85:15 (**47:48**); however, 10% decomposition to afford **49** was observed (Scheme 15).

Consequently, by increasing control over the reaction temperature the authors were able to enhance the diastereoselectivity of the reaction and remove the decomposition pathway observed in batch. The authors acknowledge that further work is required to optimize the deprotonation

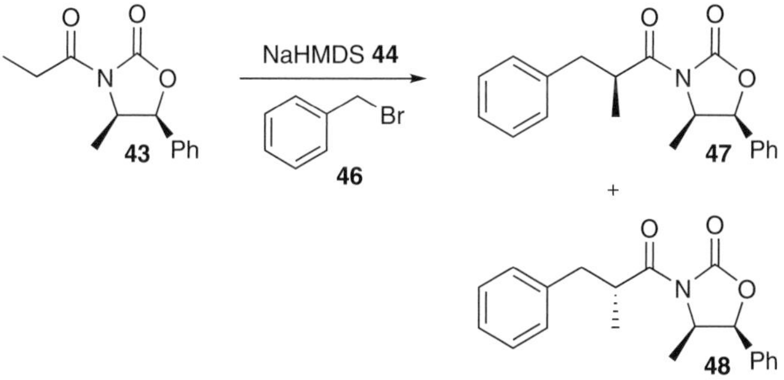

Scheme 14   The alkylation of N-acyl oxazolidinone **43** to afford diastereomers **47** and **48** investigated under continuous flow at $-100\,°C$.

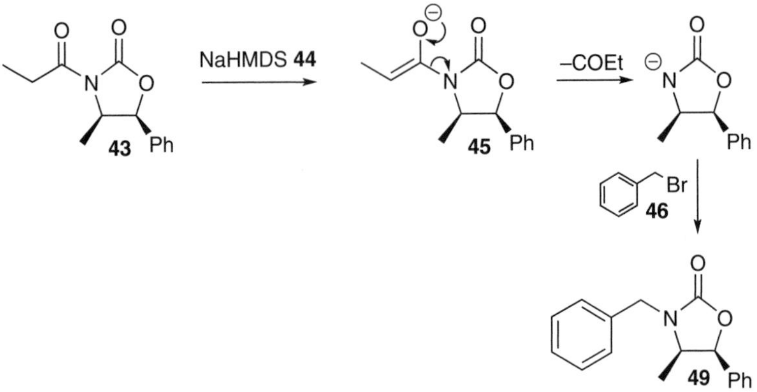

Scheme 15 The decomposition of enolate **45** observed when the reaction is conducted under batch conditions.

step in order to increase the conversion of **43** to **47/48**; however, the technique effectively demonstrated the ability to perform temperature-sensitive reactions in microfabricated reactors.

More recently, Matsuoka et al. (2006) reported the ability to supercool fluid streams within octadecylsilane-treated Pyrex microchannels, demonstrating a link between channel dimensions and the freezing point of water which range from −20 to −28 °C as the channel was reduced in width from 300 to 100 µm. Interestingly, a dimension-independent freezing temperature of −15 °C was obtained when bare Pyrex microchannels were employed. Having identified this phenomenon and found it to be independent of flow rate, the authors subsequently investigated the ability to perform asymmetric syntheses within such as system and employed the reaction depicted in Scheme 16 as a model.

Employing a biphasic reaction mixture comprising of the phase transfer catalyst (S,S)-3,4,5-trifluorophenyl-NAS bromide in aq. KOH and an organic phase containing the imine **50** and benzyl bromide **46**, the authors investigated the effect of reaction temperature on the product **51** ratio obtained. Using an untreated microchannel with dimensions of 100 µm (wide) × 40 µm (deep) × 0.6 cm (long), segmented two-phase flow was

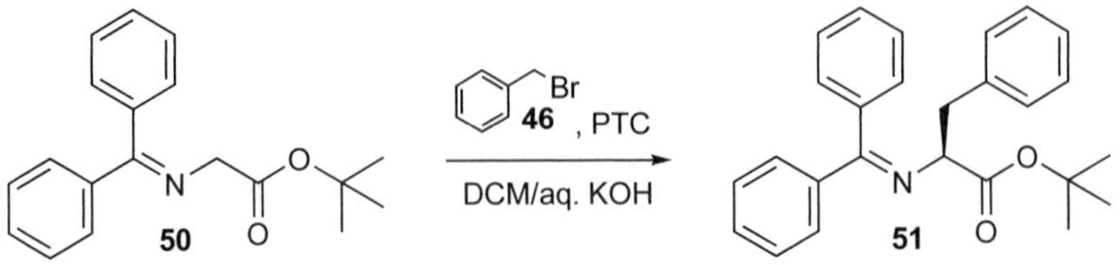

Scheme 16 Model reaction used to illustrate the application of supercooled microflows in asymmetric synthesis.

observed and the authors found that as the reaction temperature decreased from 20 to $-20\,°C$, the enantiomeric excess (*ee*) increased from 43 to >50%. Using supercooled microflow, the authors were able to evaluate the effect of reaction temperatures inaccessible within a batch reactor, whereby freezing of the aq. KOH phase was observed at $4\,°C$. Whilst the authors note that further increases in *ee* are required if the methodology is to be of widespread synthetic utility, the technique represents a means of accessing reaction conditions that are unattainable through the use of conventional bulk processes.

Another example where reactions needed to be conducted at moderately low temperatures (0–$20\,°C$) was the continuous flow synthesis of oligosaccharides reported by Carrel et al. (2007) which built on previous experience gained through the synthesis of disaccharides (Flogel et al., 2006; Geyer and Seeberger, 2007; Ratner et al., 2005). In an extension to this project, the researchers demonstrated the synthesis of a homotetramer **52** using iterative glycosylations under continuous flow (Scheme 17). In order to optimize the reaction conditions required for the glycosylation steps, the authors introduced a solution of the nucleophile into the reactor from inlet 1, glycosyl phosphate **53** (2.0 eq.) from inlet 2 and the activator, trimethylsilyl trifluoromethanesulfonate (TMSOTf) **54** (2.0 eq.), from inlet 3. Employing a deprotective quench, consisting of piperidine **55** (25% in DMF) and tetra-*n*-butylammonium fluoride (TBAF) **56**, the effect of conducting the reaction over a range of residence times (10 s to 10 min) and temperatures (0 and $20\,°C$) was investigated.

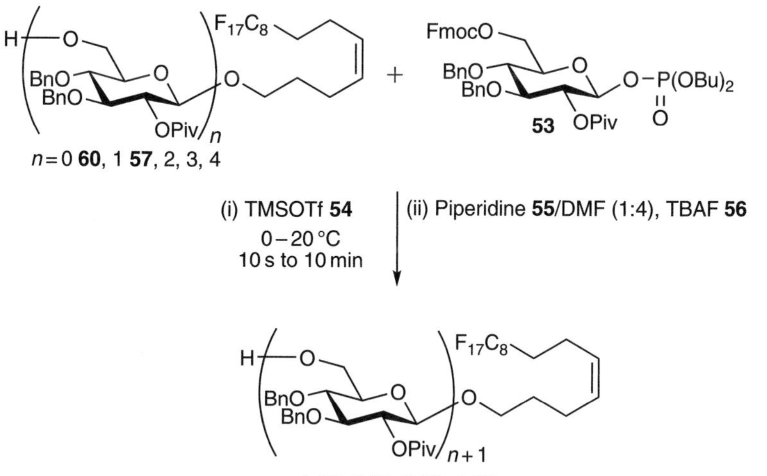

Scheme 17 General reaction scheme illustrating the iterative steps used to synthesize the β-(1 → 6)-linked D-glucopyranoside homotetramer **52**.

Using this approach, the authors identified a residence time of 30 s and a reaction temperature of 20 °C to be optimal for the synthesis of monoglycoside **57** (99% yield), which represented an improvement on conventional reaction conditions where reaction times were of the order of 30 min and temperatures typically −78 to −40 °C. After fluorous solid-phase extraction and treatment with silica gel, the monoglycoside **57** was subsequently reacted with glycosyl phosphate **53** to afford the respective disaccharide **58**, in 97% yield, with a residence time of 20 s at 20 °C. Up to this point, the residual starting material was detected within the reaction product; consequently, the authors increased the proportion of glycosyl phosphate **53** from 2.0 to 3.0 eq. and TBAF **56** from 1.5 to 2.0 eq., which resulted in complete conversion to the trisaccharide **59** (90% yield) in conjunction with a residence time of 60 s. The final step of the reaction involved the glycosylation of the previously prepared trisaccharide **59** to afford the tetrasaccharide **52**; again a residence time of 60 s was required to afford the desired product in 95% yield. Upon scaling this process, the authors successfully attained a throughput of 11.3 mmol day^{-1} for monosaccharide **57** from fluoroalkenol **60**.

2.3 The use of toxic or hazardous reagents

In addition to reactant/intermediate instability preventing the use of certain synthetic routes on a large scale, the toxicity of reagents can also precludes the use of certain transformations on a production scale. On the grounds of safety, the use of toxic or hazardous reagents and radical-based reactions are avoided due to problems associated with reaction control. As such, to date, the industry has overcome these problems through the use of alternative reagents and reaction pathways; however, *via* novel reaction methodology these reagents and techniques can be used safely and efficiently.

With this in mind, Odedra et al. (2008) evaluated the use of tris(trimethylsilyl)silane **61** as an effective reducing agent, replacing the highly toxic reagent tin hydride, for the removal of various functional groups using a glass microreactor. Building on synthetic methodology, recently reviewed by Chatgilialoglu (2008), the researchers evaluated the use tris(trimethylsilyl)silane **61** in a glass microreactor for safe and facile radical-based reductions (Scheme 18). A typical continuous flow deoxygenation/dehalogenation involved the introduction of a premixed solution of tris(trimethylsilyl)silane **61** and azobisisobutyronitrile (AIBN) **62** (1.20 M and 10 mol%, respectively) from one inlet and a solution of the substrate (1.00 M) from a second inlet. Employing a glass microreactor whereby the outlet was coupled to a back-pressure regulator (BPR), enabled the authors to superheat the reaction solvent (toluene), thus promoting the reaction without the need to employ chlorinated or toxic

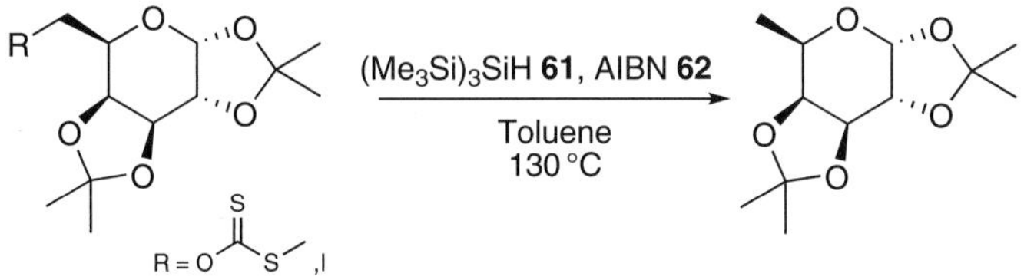

Scheme 18 Examples of the tris(trimethylsilyl)silane **61**-mediated deoxygenation and dehalogenation reactions conducted under continuous flow.

solvent systems. Using this approach the authors evaluated the generality of the flow methodology, reporting isolated yields ranging from 70 to 94% for the radical-mediated deoxygenation reactions and 67–98% for the dehalogenations.

In addition to the seven deoxygenation and dehalogenation reactions demonstrated, the authors adapted the methodology to the hydrosilylation of a series of alkynes and alkenes. To conduct such reactions, a premixed solution of the alkyne or alkene (1.00 M), tris(trimethylsilyl) silane **61** (1.2 M) and AIBN **62** (10 mol%) in toluene was passed through the heated microreactor (130 °C) at a flow rate of 200 μl min^{-1}, affording a residence time of 5 min.

As Table 5 illustrates, compared to batchwise reactions the increased heat and mass transfer obtained within the microfluidic reactor afforded enhanced *cis/trans* selectivity for reactions employing alkynes and more generically a dramatic reduction in reaction time (cf. conventional batch reactions).

Table 5 Summary of the results obtained for the continuous flow hydrosilylations using tris(trimethylsilyl)silane **61**

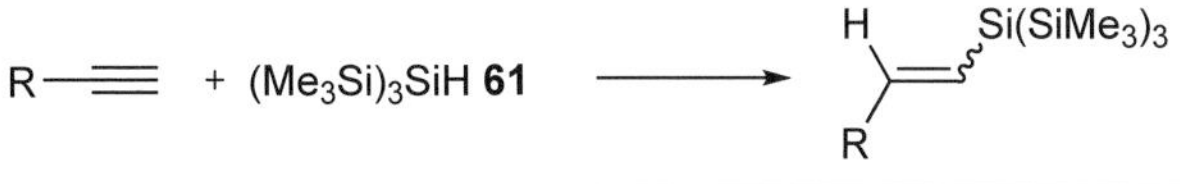

R	Z:E Ratio[a]	Yield (%)[b]
Ph **32**	98:2 (84:16)[c]	96 (88)[c]
n-C$_6$H$_{13}$	77:23	91
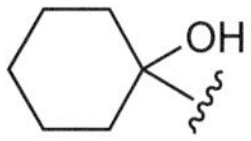 (cyclohexyl-CH(OH)-)	11:89	94

[a]Z:E ratio determined by ^{1}H NMR spectroscopy.
[b]Isolate yield.
[c]Literature value.

2.3.1 Fluorinations

With an increasing number of pharmaceutical agents containing fluorinated moieties, due to an observed enhancement in metabolic stability (cf. protonated analogs), the need for fluorinated precursors for use in drug discovery programs and subsequently in active pharmaceutical ingredient (API) production is ever rising. Owing to their rarity in nature, efficient synthetic protocols are required in order to access such fluorinated precursors; of the synthetic transformations available to the modern organic chemist, elemental fluorine (F_2) **63** and the nucleophilic fluorinating agent diethylaminosulfur trifluoride (DAST) **64** are the most extensively used. Unfortunately, DAST **64** is widely viewed as being too hazardous to employ on a large scale due to its tendency to detonate at temperatures in excess of 90 °C; as such, this powerful fluorinating agent currently fails to fulfill its synthetic potential.

Owing to the highly exothermic nature associated with converting a C–H bond to a C–F bond using elemental fluorine, the transformation has been long considered too dangerous to carry out on a large scale. However, the technique is of great synthetic utility, and as such a significant amount of research has been undertaken into conducting direct fluorination under continuous flow (Chambers et al., 2008).

In a thin-film nickel reactor, Chambers et al. (Chambers and Spink, 1999, 2001, 2005) demonstrated biphasic fluorination of a range of compounds, such as ethyl acetoacetate **65**. Employing 10% elemental fluorine **63** in a nitrogen carrier gas, the authors demonstrated a facile route to the selective preparation of the monofluorinated diketone, 2-fluoro-3-oxo-butyric acid ethyl ester **66**, with only small quantities of the product undergoing further fluorination to afford 2,2-difluoro-3-oxo-butyric acid ethyl ester **67** (Scheme 19). The group subsequently reported the fluorination of compounds such as nitrotoluene and ethyl-2-chloroacetoacetate, again demonstrating reaction control leading to excellent conversions and selectivities. As an extension to this, Chambers et al. recently published a manuscript detailing the successful scale-out of their reactor, whereby 30 reaction channels were operated in parallel. Using this approach, enabled the efficient heat transfer and gas/liquid mixing obtained in a single channel to be maintained and provided a safe alternative to hazardous batchwise fluorinations.

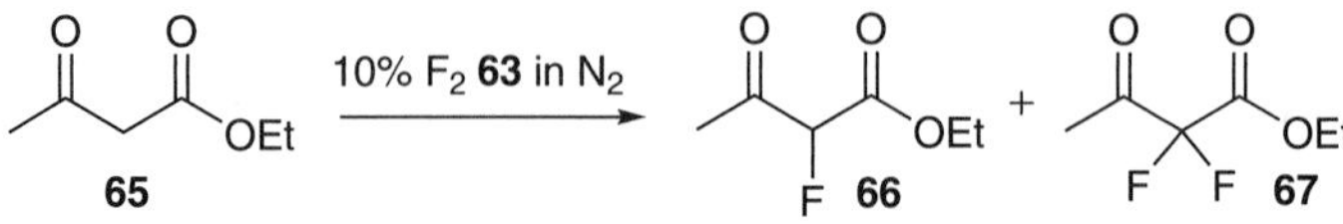

Scheme 19 Selective fluorination of a β-diketoester using elemental fluorine.

Gustafsson et al. (2008a) demonstrated a facile approach to the fluorination of a series of alcohols, carboxylic acids, aldehydes, and ketones harnessing the synthetic utility of DAST **64** in a polytetrafluoroethylene (PTFE)-based flow reactor, with an internal volume of 16.0 ml. Employing a 75 psi BPR, the authors were able to access reaction temperatures above the boiling point of the solvent under investigation. To conduct a reaction, the reactant and DAST **64** were introduced from separate inlets and mixed using a T-mixer, prior to entering the reaction channel, where a solution of sat. $NaHCO_3$ (aq.) was then introduced, after the BPR to quench the reaction mixture prior to collection of the reaction products. Upon investigating the deoxyfluorination conditions in a range of solvents, including toluene and THF, the authors rapidly identified dichloromethane (DCM) as the best solvent to use. Employing a reactor temperature of 70 °C, coupled with 1.0 eq. of DAST **64**, for the deoxyfluorination of alcohols and 2.0 eq. when using substrates containing a carbonyl moiety; the effect of reactant residence time was evaluated (8–32 min).

The authors quickly identified 16 min as being the optimum residence time and as Table 6 illustrates, under the aforementioned conditions an array of fluorinated materials were synthesized in moderate to excellent yield following an aqueous extraction and purification via column chromatography on silica gel. In addition to the array of deoxyfluorinations successfully demonstrated, the authors also evaluated the ease of performing halogen exchange, synthesizing an acyl fluoride from an acyl chloride in 80% yield.

In addition to the use of elemental fluorine **63** and DAST **64**, Miyake and Kitazume (2003) demonstrated the introduction of fluorine into a series of small organic compounds via trifluoromethylation of carbonyl compounds (Table 7), Michael addition (Scheme 20a), and Horner–Wadsworth–Emmons (Scheme 20b) reactions.

Employing a glass microreactor, with channel dimensions $= 100 \, \mu m$ (wide) $\times 40 \, \mu m$ (deep) and 8.0 cm (long), the authors investigated the trifluoromethylation of an array of aldehydes and ketones (Table 7). To perform a reaction, a solution of aldehyde/ketone (1.5 M) and trifluoromethyl(trimethyl)silane **68** (2.25 M) in THF was introduced into the microreactor from one inlet and a solution of TBAF **55** in THF from a second inlet affording a total flow rate of $1.0 \, \mu l \, min^{-1}$ and reaction time of 20 s. All microreactions were conducted at room temperature and the reaction products collected off-line in aq. HCl (1.0 N) prior to extraction of the trifluoromethylated products into diethyl ether. The resulting reaction mixtures were subsequently analyzed by ^1H and ^{19}F nuclear magnetic resonance (NMR) spectroscopy in order to determine the reaction yield. Using the aforementioned protocol, the authors were able to obtain the target trifluoromethylated alcohols in excellent yields when employing aromatic and aliphatic aldehydes (63–89% yield) as precursors; however,

Table 6 A selection of the substrates fluorinated by Gustafsson et al. (2008a) using a PTFE reactor

Substrate	Product	Yield (%)[a]
		70[b]
		61[c]
		89
		100
		100
		81
		89
		100

[a]Isolated yield.
[b]5:1 mixture of diastereomers.
[c]6:1 mixture of diastereomers.

Table 7 Summary of the results obtained for the continuous flow trifluoromethylation reactions of carbonyl-containing compounds

Substrate	Microreactor	Batch reactor
	Yield (%)[a]	Yield (%)[a]
116	89	76[b]
	83	97[b]
	63	–
	74	–
	12	–
	42	–
	7	–

[a]Yields determined by ^{19}F NMR spectroscopy.
[b]Batch reaction (5 h).

reactions of ketones afforded poor to moderate yields (7–42% yield). The authors were also pleased to find that the microreactions afforded comparable yields to those obtained under conventional batch conditions employing reaction times of 20 s compared to 5 h (Table 7).

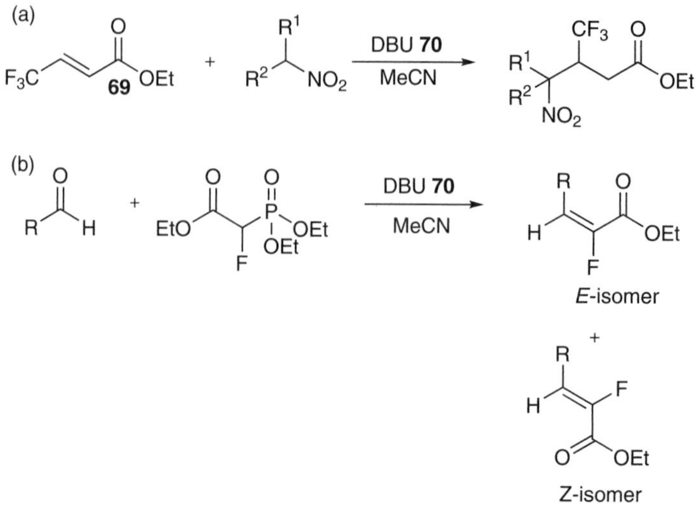

Scheme 20 A selection of the reactions investigated by Miyake and Kitazume (2003) for the incorporation of fluorine into small organic molecules (a) Michael addition and (b) Horner–Wadsworth–Emmons.

Based on previous success with the Michael addition of CF_3-containing acrylates in batch, Miyake and Kitazume (2003) and coworkers subsequently investigated the reaction under continuous flow as a means of rapidly generating a series of synthetically useful fluorinated alkanes (Scheme 20a). Again employing a reaction time of 20 s, the target compounds were obtained in isolated yields ranging from 80 to 93% which compared favorably with the results previously obtained in batch whereby yields of 90–99% were obtained with a reaction time of 30 min. In a final example, the authors reported the Horner–Wadsworth–Emmons reaction of a series of aldehydes (1.7 M) with triethyl-2-fluoro-2-phosphonoacetate **69** (2.6 M), in the presence of 1,8-dizabicycloundec-7-ene (DBU) **70**, to afford a series of α-fluoro-α,β-unsaturated esters, as illustrated in Scheme 20b. Employing dimethoxyethane (DME) as the reaction solvent and a total flow rate of $1.0\,\mu l\,min^{-1}$, the reaction products were collected off-line in aq. HCl (1.0 N) prior to extraction of the organics into diethyl ether. Upon removal of the solvent, the residue was analyzed by ^{19}F NMR spectroscopy and the yields determined using benzotrifluoride as an internal standard. The stereochemistry of the reaction products was confirmed by ^1H NMR coupling constants and the chemical shifts of the olefinic protons. Using this methodology, the authors successfully synthesized the target compounds in moderate to excellent yields; see Table 8, reporting analogous stereoselectivities to those obtained in a conventional batch reactor.

Table 8 Comparison of the stereoselectivities obtained in batch and a microreactor for the Horner–Wadsworth–Emmons reaction

R	Microreactor	Batch reactor
	Yield (%)[a]	Yield (%)[b]
C_6H_5	78	>99
	$(77:23)^c$	(70:30)
$4\text{-}CF_3C_6H_4$	88	86
	(68:32)	(64:36)
$4\text{-}MeOC_6H_4$	58	>99
	(74:26)	(76:24)
n-Nonyl	81	66
	(64:36)	(64:36)

[a]Yields determined by ^{19}F NMR spectroscopy.
[b]Isolated yields, batch reaction conducted for 30 min.
[c]The number in parentheses represent the Z:E ratio, determined by ^{19}F NMR spectroscopy.

2.3.2 Trimethylaluminum

The pyrophoric nature of trimethylaluminum ($AlMe_3$) **71** and the instability associated with the aluminum–amide intermediates make $AlMe_3$ **71** difficult to handle safely at scale, as such its use is precluded for the large-scale synthesis of amides. The ease of reactant manipulation obtained through the use of microreactors, however, affords the potential to use such reagents for the continuous flow synthesis of amides, providing a practical route to their use in large-scale production. Performing reactions on an 8.0 mmol scale, later increased to 0.2 mol, Ponten et al. (2008) demonstrated the continuous flow synthesis of an array of amides, derived from a series of methyl or ether esters and benzylamine **20**; attaining excellent isolated yields as illustrated in Table 9.

In addition to the examples provided, the authors also demonstrated the use of anilines and aliphatic amines as substrates, reporting isolated yields ranging from 65 to 98% (see Section 3), whereby this methodology is demonstrated for the synthesis of the pharmaceutically important molecules rimonabant (**72**) and efaproxiral (**73**) (Schemes 65 and 66, respectively).

2.3.3 The use of butyllithium in microreactors

The reaction of organolithium compounds with carbon-based electrophiles represents one of the most useful synthetic methodologies for the formation of C–C bonds; as such, several groups have investigated these reactions in microstructured devices.

Table 9 A selection of the trimethylaluminum-mediated reactions conducted under continuous flow

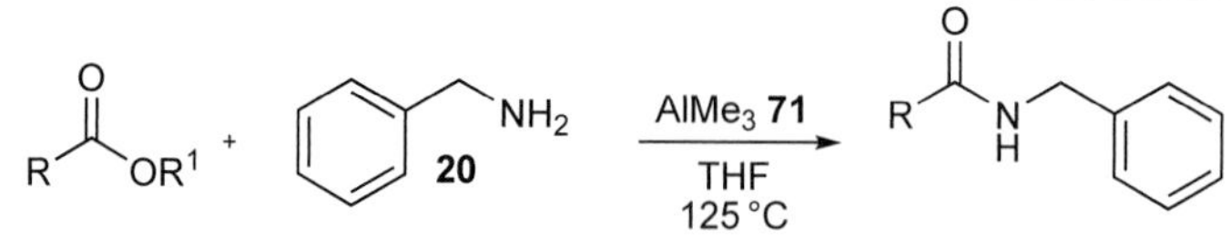

Ester	Product	Yield (%)
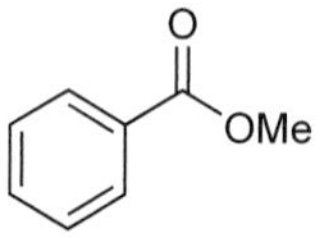	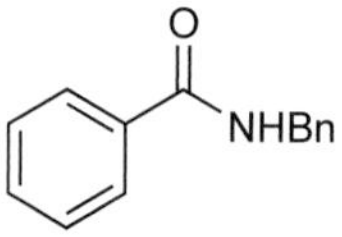	96
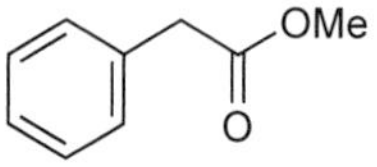	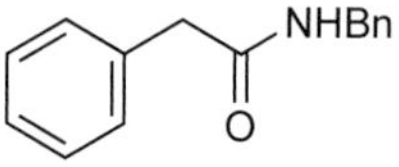	98
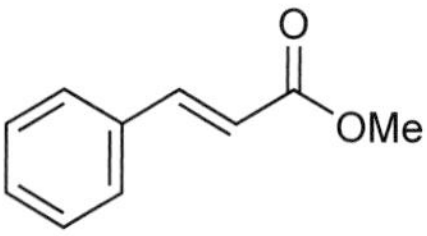	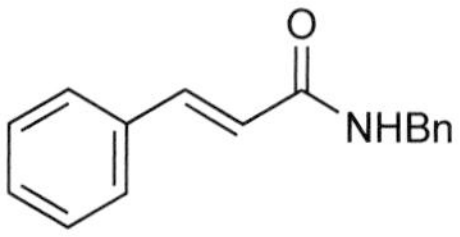	95
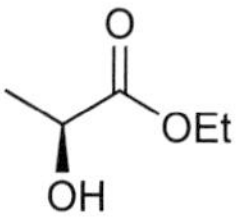	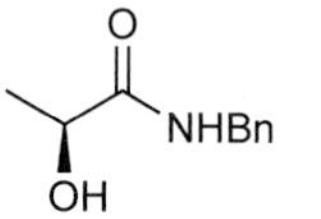	70
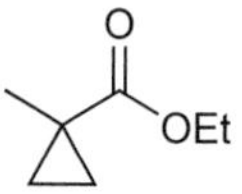	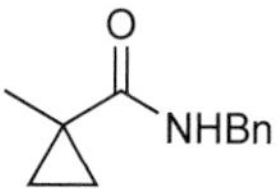	86

An early example of an investigation into the use of *n*-BuLi **74** under continuous flow conditions was demonstrated by Schwalbe et al. (2004) and involved the *in situ* generation of C_2F_5–Li **75** and its subsequent nucleophilic addition to benzophenone **76** as depicted in Scheme 21.

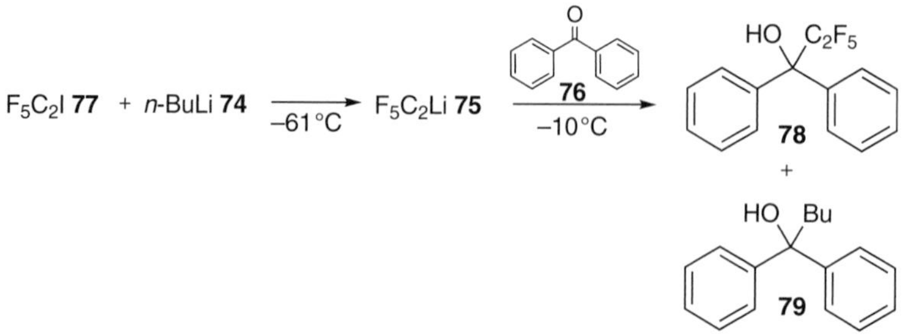

Scheme 21 Demonstration of the *in situ* generation of a short-lived organometallic species **75** and its nucleophilic addition to benzophenone **76**.

Employing a two-stage microreactor, the authors were able to conduct the first step and second steps of the reaction at different temperatures, −61 and −10 °C, respectively, using two external cooling baths. To perform a reaction, solutions of n-BuLi **74** (0.82 M in hexane) and C_2F_5I **77** (0.75 M in DCM) were pumped into the reactor at a flow rate equivalent to a residence time of 4 min. A solution of benzophenone **76** (0.62 M) in DCM was then introduced into the second reactor and a residence time of 17 min employed. The reaction products were collected prior to performing an off-line aqueous extraction and the reaction products analyzed by GC. Employing the aforementioned reaction conditions, the authors were able to obtain the target product **78** in 51% yield, representing an increase of 37% (cf. batch reactions). The authors did however report the formation of a competing side product **79** (35%) attributed to incomplete halogen–lithium exchange in the first stage of the reaction. The reaction nonetheless illustrated the use of conventionally hazardous reagents under continuous flow, along with the feasibility of conducting serial reaction steps at different temperatures.

A recent example reported by Goto et al. (2008) exploited the high rate of halogen–lithium exchange to enable the coupling of fenchone (**80**) and 2-bromopyridine **81** in a single step, compared to the more conventional two-step procedure (Scheme 22). Using this approach, it was proposed that the rapid bromine–lithium exchange could be performed chemoselectively, in the presence of a trapping agent such as a ketone. To evaluate this hypothesis, the authors selected the alkylation of the pyridinoalcohol (fenchone) (**80**) with 2-bromopyridine **81** as a model reaction and employed a stainless-steel flow reactor, comprising of a 2 ml microreactor cell and a 15 ml residence time unit. To perform a reaction, n-BuLi **74** (0.5–1.0 M) in hexane was added to a solution of fenchone (**80**) (0.5–1.0 M) and 2-bromopyridine **81** (0.46 M) in anhydrous THF at a flow rate of 5 ml min^{-1}. Maintaining the reactor at −25 °C, the authors evaluated the effect of reactant stoichiometry on the reaction at a fixed residence time of 3 min. The reaction mixture was collected in ice–H_2O to quench the reaction and the organic material extracted into ether prior to analysis by GC–MS and ^1H NMR spectroscopy.

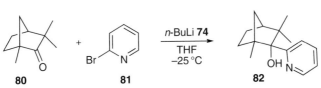

Scheme 22 Model reaction used to demonstrate the one-step coupling of fenchone (**80**) and 2-bromopyridine **81**.

Table 10 Illustration of the optimization strategy used for the one-step coupling of fenchone (**80**) and 2-bromopyridine **81** in a flow reactor

Fenchone (**80**): Halide **81**:n-BuLi **74**	Temperature (°C)	Yield (%)[a]
1:1:1	−25	56
2:1:1.5	−25	71
2:1:2	−25	91
2:1:1.5	0	68

[a]Isolated yield.

As Table 10 illustrates, using this approach the authors were able to rapidly optimize the reaction conditions, obtaining the target **82** in 91% yield when employing 2 eq. of fenchone (**80**) and n-BuLi **74**. In all cases only a single diastereomer was observed and the authors found that conducting the reaction at 0 °C resulted in a mere 3% reduction in yield. Furthermore, the reaction conditions were found to be suitable for a range of aliphatic/aromatic ketones and brominated compounds.

Nagaki et al. (2009) recently demonstrated the use of a microflow reactor for the *in situ* preparation of an oxiranyl anion derived from styrene oxide **83** and *sec*-BuLi **84**. As Table 10 illustrates, the authors evaluated the preparative synthesis of a series of substituted epoxides, by reacting 1,2-epoxyethylphenyllithium **85** with a series of electrophiles in a microflow system comprising of two T-mixers and two microtube reactors. The generation of 1,2-epoxyethylphenyllithium **85**, from styrene oxide **83** (0.10 M, 6.00 ml min^{-1}) and *sec*-BuLi **84** (0.75 M, 1.92 ml min^{-1}), was conducted in the first micromixer and was followed by trapping with MeI **86** (0.45 M, 3.84 ml min^{-1}) in the second mixer. Using this approach, the authors investigated the effect of temperature (−78 to −48 °C) and residence time (1–25 s) on the generation of intermediate **87** and subsequently the formation of 2-methyl-2-phenyloxirane.

Conducting reactions at −48 °C, the formation of oxirane occurred rapidly, indicating that the deprotonation of styrene oxide **83** is rapid; however, this was accompanied by decomposition of the oxiranyl anion **87**. Reducing the reactor temperature to −78 °C, the anion **87** was found to be stable for up to 25 s, enabling efficient reaction with a series of electrophiles to afford the respective substituted epoxide in high yield, as illustrated in Table 11.

Nagaki et al. (2008) also demonstrated the use of *sec*-BuLi **84** in a microflow system for the anionic polymerization of styrene **88**, as a means of attaining a high degree of control over the molecular weight distribution of the resulting polymer. Employing a solution of styrene **88** (2.0 M) in THF and *sec*-BuLi **84** (0.2 M) in hexane and a tubular reactor

Table 11 Summary of the reaction products generated via the deprotonation of styrene oxide **83** in a microflow reactor ($R_T = 24$ s, $-78\,°C$)

Electrophile	Product	Yield (%)[a]	Throughput (g h^{-1})
MeI **86**		88	4.2
Me$_3$SiCl		72	5.0
Benzaldehyde **116**		84[a]	6.8
Acetophenone		70[b]	6.1
Benzophenone **76**		82	9.0

[a]Diastereomer ratio 82:18.
[b]67:33, as determined by ^{1}H NMR.

[250 µm (i.d.) × 50 cm (long)], the effect of reactor temperature (24 to $-78\,°C$) on molecular weight (M_n) and molecular weight distribution (M_w/M_n) was initially investigated. As Table 12 illustrates, the M_n of the resulting polymer was directly linked to the reactor temperature and excellent M_w/M_n were obtained in all cases.

Table 12 Effect of molecular weight (M_n) and molecular weight distribution (M_w/M_n) when conducting anionic polymerizations under continuous flow

Temperature (°C)	Flow Rate (ml min^{-1})		M_n	M_w/M_n
	sec-BuLi **84**	Styrene **88**		
24	2.0	6.0	3400	1.10
0	2.0	6.0	3300	1.08
−28	2.0	6.0	3600	1.07
−48	2.0	6.0	3700	1.08
−78	2.0	6.0	4000	1.43

Employing a microreactor temperature of 0 °C, the authors demonstrated a linear relationship between the monomer **88** and initiator **84** ratio, enabling a specific molecular weight polymer to be prepared with ease. Based on these observations, the authors went on to evaluate the end functionalization of the polymers using a series of silanes such as chlorodimethylvinylsilane, with the investigation concluded with block copolymerization using functionalized polymers with different end groups. Using this approach, the authors were able to attain a high level of molecular weight distribution under relatively high reaction temperatures (0–24 °C), opening up the opportunity to prepare structurally well-defined polymers, copolymers, and block copolymers.

Using a stainless-steel micromixer coupled to a residence time unit, Tietze and Liu (2008) reported the development of a continuous flow process for the synthesis of an aminonaphthalene derivative, which is a precursor in the synthesis of novel anticancer agents. A key reaction in the compounds synthesis was the nucleophilic substitution reaction, illustrated in Scheme 23, to afford the phosphonosuccinate **89**; see Section 3 for details of the total synthesis achieved utilizing a series of microreaction steps. Prior to evaluating the use of n-BuLi **74** for the continuous flow process, the authors investigated the use of NaH **90** and organolithium bases such as lithium diisopropylamide (LDA) in batch where isolated yields of 70–91% were obtained after <24 h. The relative insolubility of the by-product NaBr precluded the use of NaH **90** in the continuous process; however, the increased solubility of LiBr meant that organo-lithium bases had potential to be employed in the system. Of the bases investigated, n-BuLi **74** was found to yield the most homogeneous solution and was subsequently employed in the microreactor, affording 70% yield of the phosphonosuccinate, ethyl 2-(diethoxyphosphoryl)acetate **89**, with a residence time of 24 min and a reactor temperature of 40 °C; representing a 13% increase in yield compared to analogous batch reactions conducted over 36 h.

2.3.4 Nitrations

The nitration of aromatics is often a rapid and exothermic process which frequently requires dropwise addition of the nitrating solution in order to prevent rapid evolution of heat, which can lead to undesirable

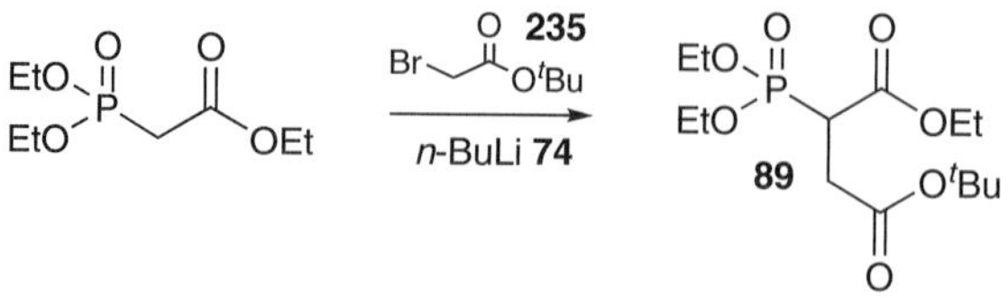

Scheme 23 Synthesis of intermediate **89** used in the preparation of gram quantities of an aminonaphthalene derivative **91**.

decomposition of the target compound and even thermal runaway. As such, the nitration of aromatics was one of the early miniaturized reactions to be reported, with Burns and Ramshaw (2002) investigating the use of a stainless-steel continuous flow reactor for the nitration of benzene **92** and toluene **93** using two-phase slugs of organic and aqueous reactants.

Ducry and Roberge (2005) subsequently reported the autocatalytic nitration of phenol **94** within a glass microreactor [$10 \times 500\,\mu m$ (wide) channels, internal volume $= 2.0\,ml$] and evaluated the effect of reactant stoichiometry on the ratio of products obtained (Scheme 24). Employing a phenol **94** solution (6%) in CH_3CO_2H (6%) and water (71%) (feed 1) and a 65% HNO_3 **95** solution (feed 2), the ratio of feedstocks was varied between 3.9:1 and 2.3:1 which corresponded to 1.1–1.8 equivalents of HNO_3 **95**. Using this approach, the effect of reaction temperature was also evaluated and it was observed that at 55 °C, autocatalysis started immediately, at 45 °C a 1 min delay was observed, and at lower temperatures, autocatalysis did not start. Through the use of a continuous flow process, the authors reported a reduction in the formation of polymeric side products (up to 70% in batch), an overall increase in product purity, and higher yields than were obtained within a batch reaction.

Employing a micromixer (T-mixer, 800 μm) and a stainless-steel reaction tube [1,380 μm (i.d.) × 12 m (long)] Kulkarni et al. (2008) investigated the aromatic nitration of salicylic acid **101** under continuous flow. Using a reactant feedstock of salicylic acid **101** in acetic acid **6** (1:16) and nitric acid **95**, reactions were conducted at 20 °C and the effect of residence time evaluated. To ensure that the results obtained were indicative of the reaction occurring within the microreactor, the reaction products were collected in a vial, stored within an ice bath, and unreacted HNO_3 **95** quenched with urea and MeOH prior to analysis of the supernatant by HPLC. Using this approach, the authors were able to prevent the undesirable decarboxylation, to 2-nitrophenol **97**, observing only the formation of the mononitrated products, 5-nitro-salicylic acid **102** (60%) and 3-nitro-salicylic acid **103** (18%) (Scheme 25). Employing increased reaction

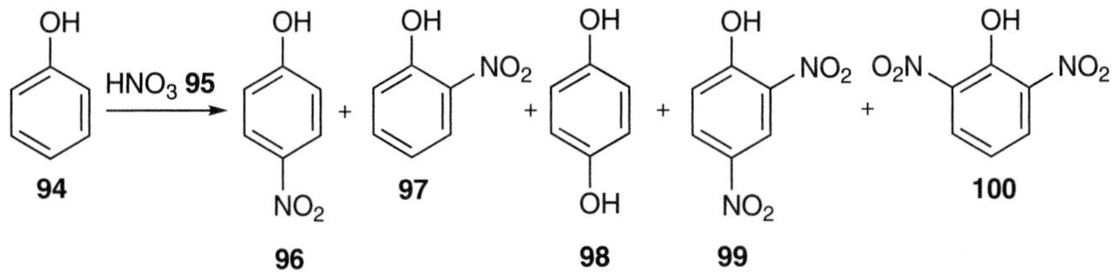

Scheme 24 Summary of the potential reaction products obtained when nitrating phenol **94**, mononitrated phenols **96** and **97**, hydroquinone **98**, and dinitrated phenols **99** and **100**.

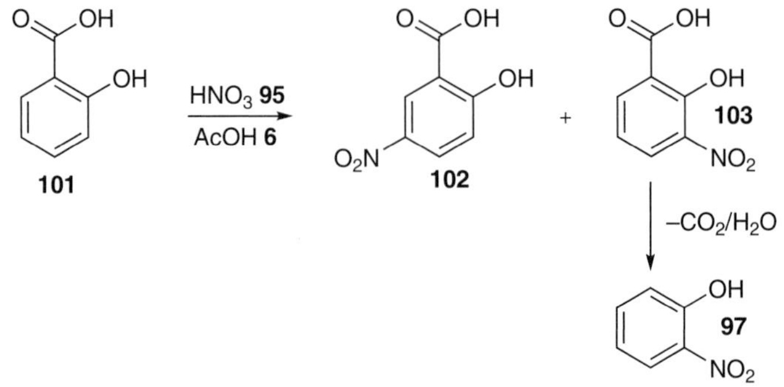

Scheme 25   Possible products obtained from the nitration of salicylic acid **101**.

temperatures the authors observed an increase in the reactions selectivity toward the synthesis of 5-nitro-salicylic acid **102**, obtaining 65% at 50 °C. The nitration of salicylic acid **101** was also the subject of a recent patent, whereby nitrations were conducted in 10–30 s, at 75 °C, within a glass microreactor (Kulkarni, 2007).

DSM recently published details of a collaboration with Corning Incorporated, which resulted in the development of a glass microreactor capable of performing a selective organic nitration, using neat HNO_3 **95**, under cGMP conditions (Scheme 26) (Braune et al., 2008). As nitration reactions are notoriously difficult to perform on a commercial scale, the main challenge was the target production volumes which were in the range of tons day^{-1}. The authors comment that although the reaction enthalpy is moderate, it is the potential exothermic decomposition of the product **104** that is problematic. Side reactions can also arise from oxidation of the starting material **105** and other components of the reaction mixture, as such careful control of the reaction conditions is required.

To perform the organic nitration, the substrate **105** and solvent were brought together in a microstructure that afforded a fine emulsion; this was followed by the addition of neat nitric acid **95** at which point the reaction started immediately. After the reaction had continued for the required time, it was stopped by the addition of water at a reduced temperature, prior to neutralization with NaOH. Neutralization was achieved in managed stages in order to control the release of heat into the reactor. Using this approach, a reactor volume of 150 ml allows the

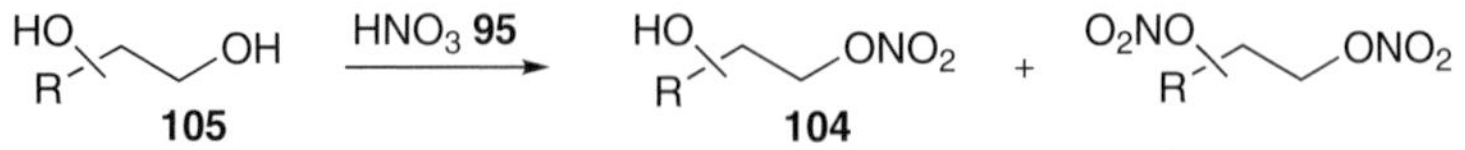

Scheme 26   Selective nitration investigated under flow by DSM using a Corning Incorporated glass reactor.

production of 13 kg h^{-1} of **104** with intrinsically high levels of safety unattainable in conventional batch techniques. Having demonstrated the viability of the microreactor, no scale-up steps were required in order to increase the throughput to meet previously discussed targets; the authors simply employed a production unit consisting of eight microreactors (two banks of four microreactors) and the unit operated under the previous conditions. Employing such a system has enabled DSM to produce 100 kg h^{-1} **104** providing an annual capacity of 800 tons if operated continuously. The investigation illustrated the ability to safely operate hazardous reactions, attain the desired chemical selectivity, design a predictable and reliable production unit, and prepare high-quality chemicals in a short time frame, all of which was achieved through the use of microreaction technology.

2.3.5 Exothermic reactions

Along with the challenges associated with the handling and use of large volumes of hazardous or toxic materials, the scale-up of exothermic reactions can also be problematic due to the difficulties associated with the efficient removal of heat generated. With this in mind, Schwalbe et al. (2004) evaluated the Paal–Knorr synthesis under continuous flow (Scheme 27). Due to the exothermic nature of the reaction, when adding ethanolamine **106** to acetylacetone **107** in batch, it was necessary to add the reagents over an extended period of time; consequently, even though the reaction was rapid, the dropwise addition of the reactants increased the reaction time considerably. In comparison, performing the reaction in a stainless-steel continuous flow reactor, with a residence time of 5.2 min, the authors were able to employ neat reactants to afford the target pyrrole **108** in 91% yield in a throughput of 260.0 g h^{-1}.

2.4 The incorporation of catalysts into microreactors

Having demonstrated the many practical advantages associated with the use of miniaturized continuous flow reactors for catalyst free, or homogeneous reactions, the following section focuses on the additional advantages that can be obtained through the use of heterogeneous catalysts, biocatalysts, reagents, and scavengers within such devices.

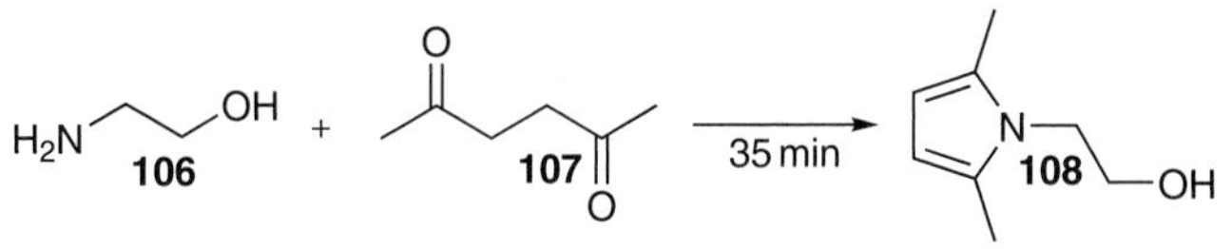

Scheme 27   Exothermic Paal–Knorr reaction conducted in a CYTOS flow reactor.

2.4.1 Base-promoted microreactions

Wiles et al. (2004a, 2007a) demonstrated the incorporation of a series of silica-supported bases into an EOF-based capillary reactor, as a means of increasing the efficiency of the Knoevenagel condensation (cf. conventional batchwise protocols). To conduct a reaction, the solid-supported base was packed into a borosilicate glass capillary [500 μm (i.d.) × 3.0 cm (long)] and held in place by two microporous silica frits. A premixed solution of the aldehyde and ethyl cyanoacetate **109**/malononitrile **110** (1.00 M in MeCN, respectively) was then mobilized through the packed bed, where the base-catalyzed condensation reaction occurred, and the reaction products collected at the reactor outlet in MeCN, prior to off-line analysis by GC–MS. Using this approach, the authors investigated the reactivity of a range of substituted aromatic aldehydes, obtaining conversions in the range of 99–100% and solid-supported catalysts **111–114** (Scheme 28). Compared to traditional stirred or shaken reactors, the use of a continuous flow system proved advantageous as it led to reduced degradation of the solid-supported catalyst, leading to enhanced reagent lifetimes and between run reproducibility.

Having demonstrated the principle using small quantities of catalytic material, typically 5 mg, the authors later demonstrated the ability to generate the aforementioned materials with a throughput ranging from 0.57 to 0.94 g h^{-1} by simply increasing the size of the packed bed employed [3,000 μm (i.d.) × 3 cm (long)].

Employing silica-supported piperazine **113** (0.10 g, 1.70 mmol N g^{-1}), the authors demonstrated the use of EOF for the semipreparative scale synthesis of a series of α,β-unsaturated compounds and as previously observed in the capillary-based reactor, the compounds were obtained in excellent isolated yield and product purity (Table 13).

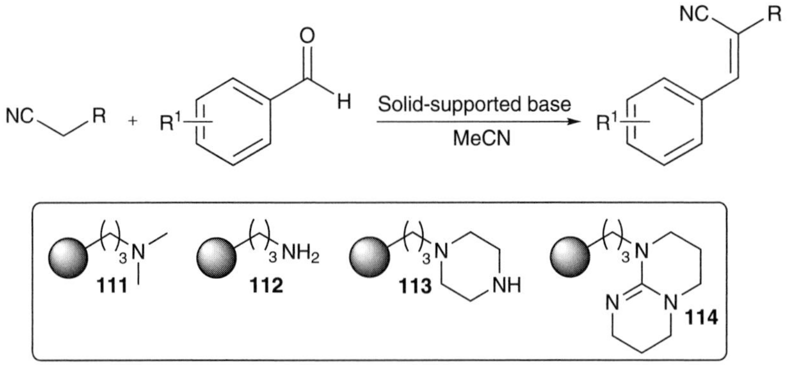

Scheme 28 General reaction scheme illustrating the Knoevenagel condensation conducted using EOF as a pumping mechanism.

Table 13 A selection of the α,β-unsaturated compounds synthesized using a silica-supported base **113** and EOF as the pumping mechanism

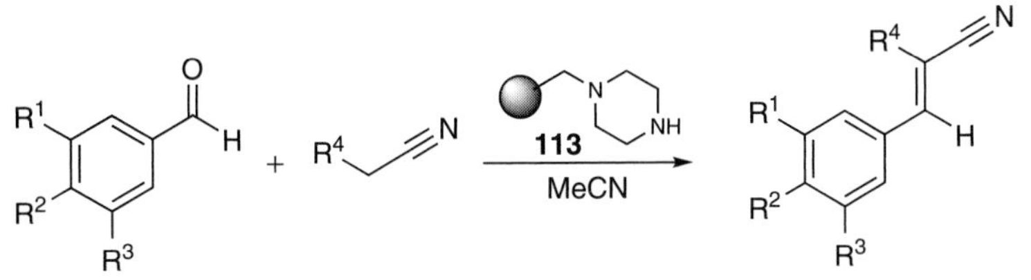

R¹	R²	R³	R⁴	Flow rate ($\mu l\,min^{-1}$)	Yield (g)[a]	Yield (%)
H	H	H	$CO_2C_2H_5$	62.0	0.75	99.7
H	CO_2CH_3	H	$CO_2C_2H_5$	56.1	0.87	99.8
OCH_3	H	OCH_3	$CO_2C_2H_5$	50.1	0.78	99.9
H	OBn	H	$CO_2C_2H_5$	51.1	0.94	99.7
H	Br	H	$CO_2C_2H_5$	55.1	2.30[b]	99.4
H	H	H	CN	62.1	0.57	99.4
H	CO_2CH_3	H	CN	60.4	0.76	98.8
OCH_3	H	OCH_3	CN	55.7	0.71	99.2
H	OBn	H	CN	48.4	0.75	99.4
H	Br	H	CN	48.3	1.70[b]	99.8

[a]Unless otherwise stated, reactions were conducted for 1.0 h.
[b]Reaction conducted for 2.5 h.

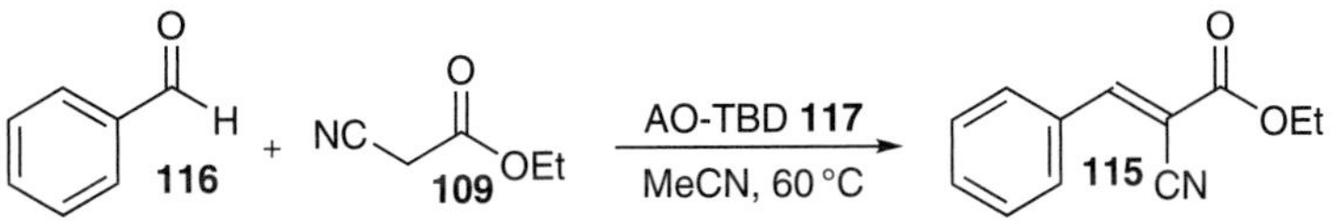

Scheme 29 Model reaction employed by McQuade et al. for the evaluation of an immobilized base **117**.

Bogdan et al. (2007) subsequently investigated the synthesis of (E)-ethyl-2-cyano-3-phenylacrylate **115** (Scheme 29) via the base-catalyzed condensation of ethyl cyanoacetate **109** and benzaldehyde **116** in a tubular reactor [1.6 mm (i.d.) × 10–60 cm (length)] packed with polymer-supported 1,5,7-triazabicyclo[4.4.0]undec-3-ene **117**. Employing a premixed solution of ethyl cyanoacetate **109** and benzaldehyde **116** (0.43 and 0.39 M, respectively), the reactor was heated to 60 °C using a HPLC column heater and the effect of residence time evaluated (25–300 s). Using the aforementioned procedure, the authors found a residence time of 300 s afforded the target compound **115** in 93% conversion and a throughput of $0.2\,g\,h^{-1}$.

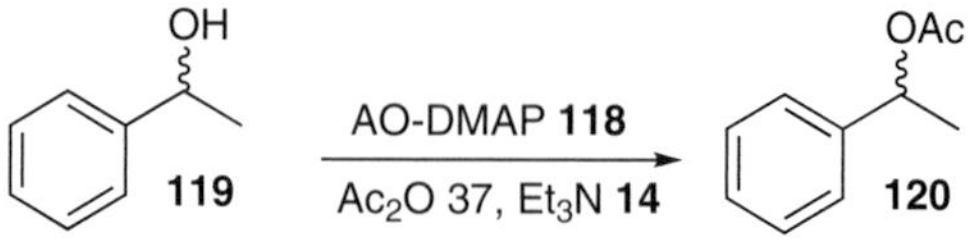

Scheme 30 Evaluation of AO-DMAP **118** as a catalyst for acylation of 2° alcohols.

Having proved the synthetic utility of their system, the authors subsequently evaluated a second supported base, 4-dimethylaminopyridine (AO-DMAP) **118**, toward the acylation of 2° alcohols (Scheme 30). Employing a premixed solution of phenyl-1-ethanol **119**, Et$_3$N **14** and acetic anhydride **37** (0.33, 0.50, and 0.50 M) in hexane, reactions were conducted at room temperature and the effect of residence time evaluated (10–50 s). Using a 60 cm packed bed, the authors were able to obtain near quantitative conversions to **120** employing residence times <20 s, with flow reactions providing superior results to those obtained in analogous batch reactions.

More recently, Costantini et al. (2009) reported a novel technique for the incorporation of an organic base, 1,5,7-triazabicyclo[4.4.0]dec-5-ene (TBD), into silicon microchannels [100 μm (wide) × 100 μm (deep) × 1.03 m (length)] which involved derivatization of a PGMA polymer brush wall coating (150 nm). To assess the catalytic properties of the derivatized polymer brushes, the authors employed the Knoevenagel condensation as a model reaction, using MeCN as the reaction solvent and a reaction temperature of 65 °C (Scheme 31). Using this approach, the authors were able to produce the target condensation product **121** with an hourly output of 7.5×10^{-5} M (using only 2.9×10^{-5} mmol of TBD **122**), with no leaching of the base observed.

Baxendale et al. (2008) reported a bifurcated approach to the synthesis of thiazoles and imidazoles by coupling a glass microreactor and a packed-bed reactor to achieve a base-mediated condensation reaction. As Scheme 32 illustrates, reactions focused on the use of ethyl isocyanoacetate **123**, as the cyanide source, with variations made via the isothiocyanate reagent, as illustrated in Table 13.

Initial investigations employed a reaction temperature of 55 °C and a flow rate of 100 μl min^{-1} with ethyl isocyanoacetate **123** (0.75 M)

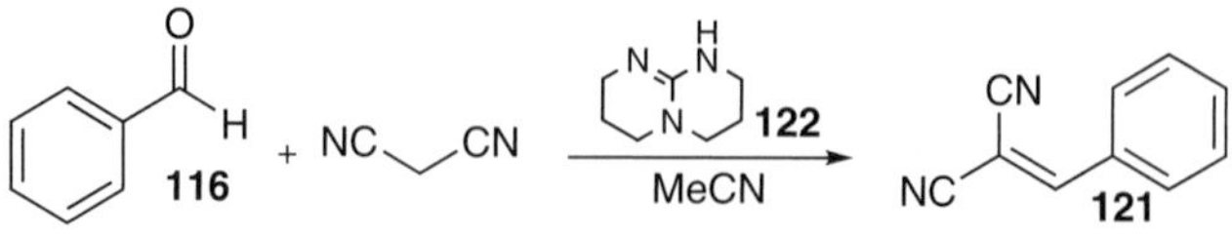

Scheme 31 Model reaction used to demonstrate the catalytic efficiency of derivatized polymer bushes.

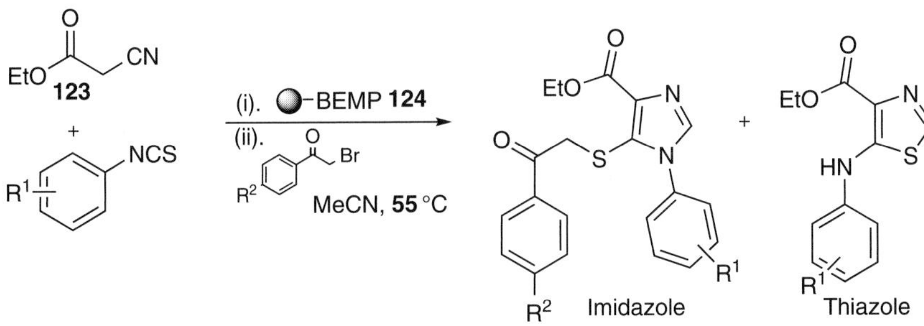

Scheme 32 Schematic illustrating the general reaction conditions employed for the continuous flow synthesis of thiazoles and imidazoles.

4-bromophenylisothiocyanate (0.75 M) reacted in the presence of 2-*tert*-butylimino-2-diethylamino-1,3-dimethylperhydro-1,3,2-diazaphosphorine on polystyrene (PS-BEMP) **124** to afford the target thiazole in excellent purity (95%) but only 58% yield. Assuming that the reminder of the thiazole was trapped on the polymer-supported base **124**, the authors passed a solution of electrophile (0.75 M) through the packed bed to afford a new product, the respective imidazole (38%). Optimal conditions were found to be 1:1 ratio of coupling reagents (0.75 M in MeCN), 1.6 eq. of PS-BEMP **124**, and a flow rate of 50 µl min^{-1} (for each reactant solution employed). Using the aforementioned conditions, the authors evaluated the effect of changing the functionality on the isothiocyanate precursor and as summarized in Table 14, an electronic effect can be seen with methoxy substituents favoring the selective formation of the thiazole.

Table 14 Summary of the results obtained for the continuous flow synthesis of thiazoles and imidazoles

R^1	R^2	Isolated yield		Combined yield (%)
		Thiazole (%)	Imidazole (%)	
4-Cl	OMe	47.5	47.5	95.1
4-Br	Br	53.0	30.0	83.0
4-OMe	H	96.0	4.0	100.0
3-F	Br	68.0	28.0	96.0
3-OMe	Br	84.0	5.0	89.0
2-OMe	H	90.0	7.0	97.0
3,5-CF$_3$	OMe	83.5	10.5	94.0
3,4-Cl	CN	53.0	26.0	79.0

2.4.2 Acid-catalyzed microreactions

Using a similar reaction setup to that reported for the Knoevenagel condensation, Wiles et al. (2005) investigated the acid-catalyzed protection of aldehydes as their respective dimethyl acetal (Scheme 33). Again, reactions were performed by pumping a premixed solution of benzaldehyde 116 and trimethylorthoformate 125 (1.0 and 2.5 M, respectively, in MeCN) through the packed-bed reactor containing Amberlyst-15 126 (5.0×10^{-3} g, 4.20 mmol g^{-1}). Off-line analysis of the reaction products by GC–MS was then used to determine the conversion of aldehyde 116 to acetal 127 and the flow rate adjusted accordingly. Using this approach, dimethoxymethyl benzene 127 was obtained in 99.8% yield (% RSD = 0.13, $n = 5$) and a throughput of 2.5×10^{-2} g h^{-1}. The authors subsequently investigated the synthesis of a series of dimethyl acetals (Table 15), all of which were obtained in excellent yield (>95.4%) and purity, without the need for additional off-line purification.

Using the same reactor, a premixed solution of aldehyde or ketone and dithiol (1.0 M, 1:1) in anhydrous MeCN was passed through the Amberlyst-15 126 containing packed bed. The reaction products were again collected in MeCN and analyzed every 10 min by GC–MS. Once optimized, the reactions were conducted for 1 h, to produce the required quantity of each 1,3-dithiane or 1,3-dithiolane, the reaction products were then concentrated *in vacuo* and analyzed by ^{1}H NMR spectroscopy (Tables 16 and 17) (Wiles et al., 2007b).

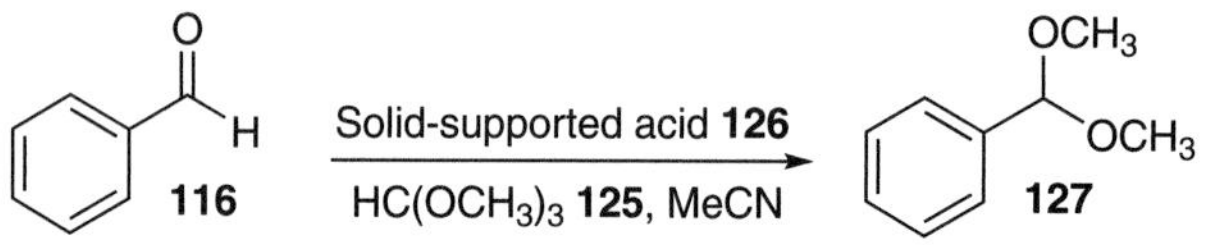

Scheme 33 An example of the protection of an aldehyde 116 as its dimethyl acetal 127 performed using a solid-supported acid catalyst under EOF-driven continuous flow.

Table 15 A selection of the results obtained for the synthesis of dimethyl acetals under continuous flow

Aldehyde	Flow rate (μl min^{-1})	Yield (%)
Benzaldehyde 116	1.75	99.8
4-Bromobenzaldehyde 134	1.00	99.9
4-Chlorobenzaldehyde	1.60	99.8
4-Cyanobenzaldehyde	2.00	99.6
2-Naphthaldehyde	1.40	99.8
Methyl-4-formylbenzoate	0.60	99.9
3,5-Dimethoxybenzaldehyde 268	0.50	98.8

Table 16   A selection of the continuous flow thioacetalizations reported by Wiles and coworkers (2004a)

Aldehyde	n	Flow rate ($\mu l\,min^{-1}$)	Yield (%)
Benzaldehyde **116**	2	63.7	99.97
	1	63.4	99.97
4-Bromobenzaldehyde **134**	2	61.4	99.92
	1	61.2	99.96
4-Chlorobenzaldehyde	2	61.7	99.91
	1	61.9	99.95
4-Cyanobenzaldehyde	2	65.4	99.94
	1	64.6	99.96
4-Benzyloxybenzaldehyde	2	61.1	99.22
	1	60.9	99.93
4-Methylbenzaldehyde	2	69.7	99.97
	1	69	99.93
4-Biphenylcarboxaldehyde	2	63	99.06
	1	63	99.97
2-Naphthaldehyde	2	60.4	99.94
	1	60.2	99.98
2-Furaldehyde	2	67.9	99.92
	1	67.5	99.97
3,5-Dimethoxybenzaldehyde **268**	2	67.9	99.91
	1	67.7	99.93

In addition to developing a facile route to the synthesis of thioacetals and thioketals, optimization of the reagents residence time within the packed bed enabled the authors to demonstrate the chemoselective protection of aldehydic functionalities in the presence of ketonic moieties (Scheme 34).

Compared to batch protocols, the flow reaction afforded the target compound **128** in quantitative yield and purity, with no sign of the competing ketal **129** formation. Although the flow reactions proceeded to near quantitative conversion in all cases, should the 1,3-dithiane or 1,3-dithiolane be used in subsequent reaction steps where Pd catalysts are employed for example, it is crucial than no dithiol residues are present. With this in mind, the authors packed a plug of silica gel impregnated

Table 17   A summary of the continuous flow thioketalizations performed using EOF

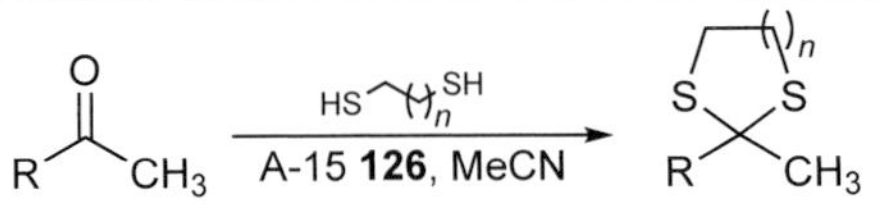

Ketone	n	Flow rate ($\mu l\, min^{-1}$)	Yield (%)
Acetophenone	2	41.5	99.57
	1	41.3	99.96
Propiophenone	2	40.2	99.97
	1	40.3	99.96
Butyrophenone	2	41.6	99.9
	1	41.6	99.9
Cyclohexanone	2	42.2	99.62
	1	42.1	99.98
Benzophenone **76**	2	40.2	99.81
	1	40.1	99.91
4-Nitroacetophenone	2	40.9	99.95
	1	41	99.95
2-Methoxyacetophenone	2	40.9	99.93
	1	41.9	99.93
4-Chloroacetophenone	2	40.9	99.87
	1	40.8	99.91
4-Hydroxyacetophenone	2	42.2	99.76
	1	42.1	99.83
4-Bromoacetophenone	2	40.2	99.94
	1	40.1	99.97

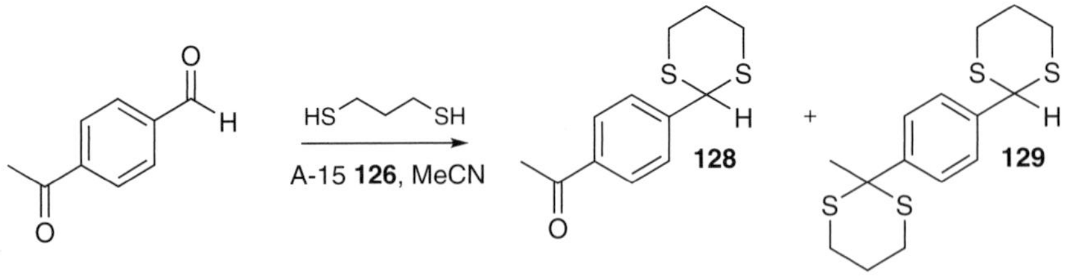

Scheme 34   Potential reaction products attainable in the protection of bifunctional compounds.

with copper sulfate into the flow reactor in order to scavenge the $<2 \times 10^{-3}\%$ dithiol present in the reaction mixture. As the scavenger turned from blue to yellow upon saturation, exhaustion of the scavenger was self-indicating.

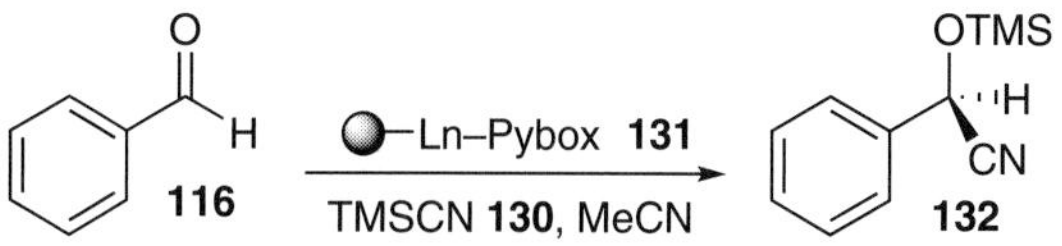

Scheme 35   An example of enantioselective synthesis performed in a packed-bed microreactor.

Another example of a Lewis acid-catalyzed reaction conducted in a miniaturized packed-bed reactor was demonstrated by Lundgren et al. (2004). As Scheme 35 illustrates, the model reaction selected involved the enantioselective addition of trimethylsilyl cyanide (TMSCN) **130** to benzaldehyde **116**. Microreactions were conducted by pumping a premixed solution of benzaldehyde **116** and TMSCN **130**, in MeCN, through a packed bed containing a polymer-supported lanthanide–pybox catalyst **131** under EOF conditions. The reaction products were diluted upon collection with MeCN and analyzed chromatographically, off-line, in order to determine both the conversion and enantioselectivity of the cyanohydrin prepared. Compared to standard batch techniques, this approach proved advantageous as it enabled the enantioselective synthesis of cyanohydrins **132**, without the need for additional extraction steps in order to recover, and potentially reuse, the catalyst from solution. In the same year, the authors also communicated the results of an investigation into the use of a series of additives, this time employing a homogeneous catalyst, for the same model reaction.

More recently Wiles and Watts (2008b, 2008c) reported the fabrication of a borosilicate glass microreactor capable of performing solution-phase syntheses, followed by heterogeneously catalyzed reaction steps and utilized this approach for the multicomponent synthesis of 51 α-aminonitriles. Initial investigations were conducted using a polymer-supported ethylenediaminetetraacetic acid ruthenium (III) chloride (PS-RuCl$_3$) **133** as the catalyst (10 mg, 0.26 mmol g^{-1}). To perform a reaction, the authors reacted a range of aromatic and aliphatic amines (0.4 M in MeCN) with 4-bromobenzaldehyde **134** (0.4 M in MeCN) in a microchannel [150 μm (wide) × 50 μm (deep) × 5.6 cm (long)], the *in situ* synthesized imine (0.2 M) was then mixed with TMSCN **130** (0.2 M in MeCN) prior to passing through the catalyst bed, where the nucleophilic cyanide addition occurred to afford the target α-aminonitrile with throughputs ranging from 17.2 and 25.4 mg h^{-1} (Table 18).

In order to increase the throughput of the system, the authors subsequently investigated the use of an alternative catalyst, polymer-supported scandium triflate (PS-Sc(OTf)$_2$) **135**. As Table 18 illustrates, compared to PS-RuCl$_3$ **133**, the PS-Sc(OTf)$_2$ **135** was found to be a more active catalyst toward the Strecker reaction and afforded the target α-aminonitriles in

Table 18   Comparison of PS-RuCl$_3$ **133** and PS-Sc(OTf)$_2$ **135** catalysts toward the Strecker reaction, conducted under continuous flow

Product	PS-Catalyst	Throughput (mg h^{-1})
	133	17.2
	135	34.4
	133	18.1
	135	36.1
	133	18.9
	135	38.0
	133	19.7
	135	39.5
	133	32
	135	63.6

34.4–63.6 mg h^{-1}. Having demonstrated the ability to synthesize a series of pharmaceutically interesting α-aminonitriles under continuous flow, the authors used the methodology to prepare a library of 50 compounds, constructed from 10 substituted aldehydes and 5 amines, whereby excellent yields and purities were obtained in all cases.

2.4.3 Metal-catalyzed reactions

As a means of increasing the efficiency of a series of commonly employed C–C cross coupling reactions, including the Suzuki–Miyaura and Heck–Mizoroki, Mennecke et al. (2008) developed an oxime-based palladacycle catalyst and evaluated its coordinative immobilization within a PASSflow (polymer-assisted solution-phase synthesis under flow conditions) reactor. Employing a flow reactor comprising of (polyvinyl) pyridine-coated Raschig rings, functionalized with a palladacycle **136** (10 mmol Pd ring^{-1}), the authors initially evaluated the Suzuki–Miyaura reaction, as illustrated in Table 19. The reactions were conducted by circulating a solution of aryl bromide (1.00 mmol), boronic acid (1.50 mmol), and cesium fluoride **137** (2.4 mmol) in DMF–H$_2$O (10:1, 5 ml), through the heated reactor (100 °C) at a flow rate of 2.5 ml min^{-1}. After a period of 24 h, the reactor was rinsed with DMF–H$_2$O and the washings diluted with H$_2$O prior to extraction with ethyl acetate. The resulting reaction products were subsequently purified by flash chromatography on silica gel to afford the target substituted biphenyl in moderate to excellent yield.

In addition to the Suzuki–Miyaura reaction, the authors also found the catalyst **136** to be active toward the arylation of olefins with aryl halides as illustrated in Table 20. Again to conduct a reaction, the authors circulated a solution containing 4-iodoacetophenone **138** (1.00 mmol), the alkene (3.00 mmol) under investigation, and tributylamine **139** (3.00 mmol) in anhydrous DMF (3 ml), through the heated reactor (120 °C) at a flow rate of 2 ml min^{-1}. After 24 h, the flow reactor was rinsed with DMF and the

Table 19 Summary of the results obtained for the continuous flow Suzuki–Miyaura reaction using an immobilized Pd (II) catalyst **136**

R^1	Residence time (h)	Yield (%)
H	24	89
CH$_3$	24	56
COCH$_3$	9	91
OCH$_3$	24	50

Table 20 A selection of the results obtained for the Heck-Mizoriki reaction under continuous flow

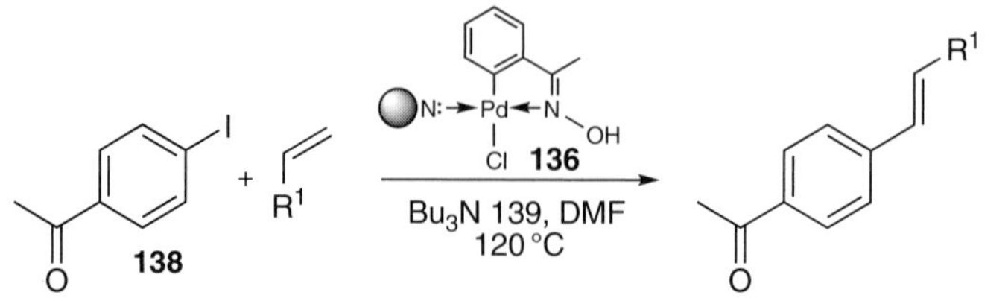

R^1	Residence time (h)	Yield (%)[a]
CO_2Cy	2	99
CO_2t-Bu	4	99
CN	4	97 (5:1)[b]

[a]Unless otherwise stated, only the E isomer was isolated.
[b]The number in parentheses represents the E/Z ratio.

reaction products diluted with water prior to extraction into ethyl acetate, to afford the target product in excellent purity.

Using an immobilized salen complex **140**, Annis and Jacobsen (1999) investigated the hydrolytic kinetic resolution of *rac*-4-hydroxy-butene oxide **141**. By simply pumping a solution racemate **141** through a column containing the silica-supported catalyst **140**, the authors were able to generate the desired triol **142** in excellent *ee*, as summarized in Scheme 36. The approach was found to be advantageous compared to those resolutions conducted in analogous batch reactions, as the technique negates the need for laborious purifications in order to isolate the product and recover the catalyst.

In addition to employing packed beds, a small group of researchers have demonstrated the use of wall coating as a means of incorporating heterogeneous catalysts into continuous flow systems. One such example

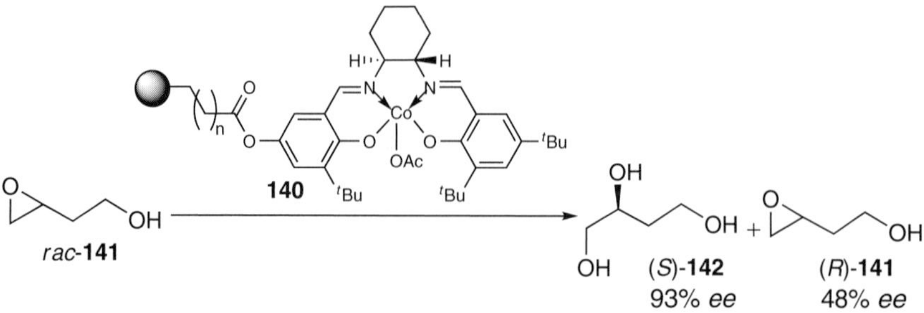

Scheme 36 Kinetic resolution of *rac*-4-hdroxy-1-butene oxide **141**.

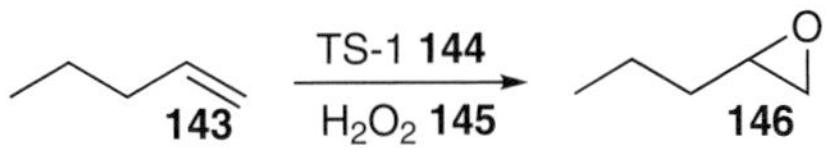

Scheme 37   The use of a wall-coated microreactor in the epoxidation of 1-pentene **143**.

was the epoxidation of 1-pentene **143**, reported by Wan et al. (2002), which was achieved using a microchannel coated with a zeolite layer **144** (Scheme 37). Coating a silicon microchannel [500 μm or 1,000 μm (wide) 250 μm (deep)] with a 3-μm-thick layer of TS-1 zeolite **144**, the oxidation of 1-pentene **143** was investigated in the presence of hydrogen peroxide **145** and afforded epoxypentane **146**. Reducing the channel width to 500 μm, the authors were able to double the conversion to **146** an observation that is attributed to an increase in the surface to volume ratio.

In addition to packed and wall-coated systems, numerous researchers have investigated the fabrication of membranes, within microchannels, in which catalytic material can be incorporated. Employing a protocol developed by Kenis et al. (1999), Uozumi et al. (2006) deposited a poly(acrylamide)-triarylphosphane palladium membrane (PA-TAP-Pd) (1.3 μm (wide), 0.37 mmol g^{-1} Pd) within a glass microchannel [100 μm (wide) × 40 μm (deep) × 1.4 cm (long)]. Once formed, the membrane was used to catalyze a series of Suzuki–Miyaura C–C bond-forming reactions, the results of which are summarized in Table 21.

To perform a reaction, the authors employed two reactant streams, the first contained the aryl iodide (6.3×10^{-3} M) in EtOAc/2-PrOH (1:2.5) and

Table 21   Summary of the results obtained for the Suzuki-Miyaura coupling reactions conducted using an *in situ* fabricated PA-TAP-Pd membrane

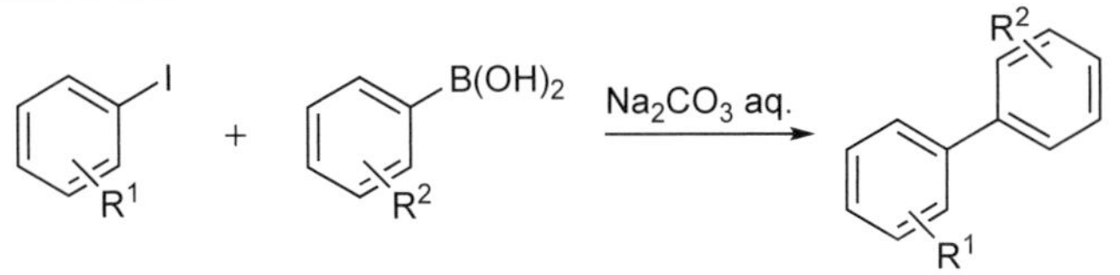

R^1	R^2	Yield (%)
H	4-MeO	99
H	3-Me	96
H	2-Me	99
3-EtCO$_2$	4-Me	99
3-Cl	4-MeO	88
4-CF$_3$	4-MeO	99

the second the aryl boronic acid $(9.4 \times 10^{-3} \text{M})$ in aq. Na_2CO_3 $(1.8 \times 10^{-2} \text{M})$. The reagents were infused through the heated reactor $(50\,°C)$ at a flow rate of $2.5\,\mu l\,min^{-1}$, which afforded a residence time of 4 s, and the biphasic reaction products collected prior to off-line analysis by GC and ^1H NMR spectroscopy. As Table 21 illustrates, in all cases excellent yields were obtained, ranging from 88 to 99%, demonstrating the high catalytic activity of the PA-TAP-Pd membrane. Importantly, in the absence of the membrane no coupling products were obtained.

Within the pharmaceutical industry, another important part of a reaction process is product purification and although the use of microreaction technology has provided overwhelming evidence to show that reactions are more efficient and products are synthesized in higher purity (cf. batch procedures), trace metal contamination still presents a problem. Hinchcliffe et al. (2007) recently developed a reactor capable of removing trace metal contaminants, such as Pd, Co, Cu, and Hg, from reaction mixtures under continuous flow. Using the Pd-catalyzed Suzuki reaction as a model, the authors screened a series of commercially available solid-supported scavengers, ranging from silica gel, carbon, and various QuadraPure^{TM} resins (thiourea, iminodiacetate and aminomethyl phosphonic acid) for the removal of Pd at a concentration of 60 ppm. Of the materials evaluated, QuadraPure^{TM} thiourea-derived was found to be the best scavenger, removing >99% of Pd from the reaction mixture in a single pass. Coupled with the fact that the proportion of trace metal contaminants detected within continuous flow reaction products are inherently low, due to reduced catalysts degradation, the use of a scavenger cartridge at the end of a reaction sequence presents a relatively long-term solution to this problem.

2.4.4 Multiple catalyst systems

Using a series of PASSflow reactors, with typical dimensions of 5 mm (i.d.) × 10 cm (long), Kirschning et al. (Kirschning and Gas 2003; Kirschning et al., 2001, 2006) demonstrated the ability to combine a series of immobilized reagents in order to perform a sequence of reaction steps. In one example, the group demonstrated an oxidation, followed by a silyl deprotection and reductive amination steps to afford a derivatized steroid **147**, as illustrated in Scheme 38, in excellent yield and purity.

In 2007, Wiles et al. (2007c) demonstrated the ability to employ solid-supported catalysts in series within an EOF-based microreactor. As summarized in Scheme 39, the model reaction sequence involved combining a previously investigated acid-catalyzed deprotection with a base-catalyzed condensation reaction to enable the synthesis of α,β-unsaturated carbonyl compounds from dimethyl acetals.

The microreactor comprised of a borosilicate glass capillary [500 μm (i.d.) × 3.0 cm (long)], connected to two reservoirs, one contained the

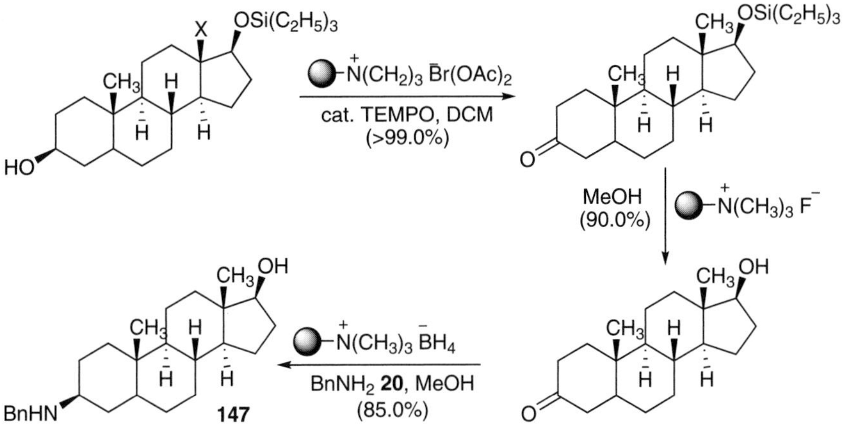

Scheme 38 Polymer-assisted derivatization of a steroid under continuous flow conditions.

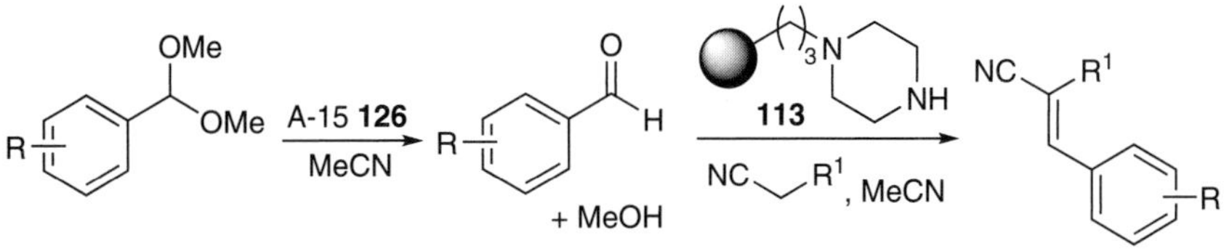

Scheme 39 General reaction scheme illustrating the multistep synthesis of an α,β-unsaturated compound.

reactant solution and the other served to collect the reaction products. To retain the solid-supported catalysts, a microporous silica (MPS) frit was placed at one end of the capillary and Amberlyst-15 **126** (2.5 mg) dry packed against it, a second frit followed and the solid-supported base **113** (2.5 mg) was packed into the reactor and held in place by a third MPS frit; this way the catalysts remained spatially resolved.

A typical procedure for the microreactions involved passing a solution of dimethyl acetal and activated methylene (1.0 M, respectively) in MeCN, through the solid-supported acid, where deprotection to afford the aldehyde occurred, followed by the supported base, where the *in situ* generated aldehyde condensed with the activated methylene to afford the desired α,β-unsaturated compound. Using EOF as the pumping technique, the reaction products were collected in MeCN, analyzed off-line by GC–MS and the conversion of starting materials to product quantified. Should any starting materials remain, the flow rate was reduced, by reducing the applied field, and the reaction repeated until optimized. Once successfully optimized, the reactions were operated continuously for 2.5 h, the reaction solvent was then removed *in vacuo* and the "crude" product analyzed by ^{1}H NMR spectroscopy. Where compounds had not previously been reported within the literature, elemental analysis was performed using the product generated under flow conditions. Using this approach, the authors synthesized 20 α,β-unsaturated compounds in isolated yields of 99.1–99.9%, >99.9% purity utilizing residence times in the range of 0.6–1.25 min, depending on the reactants employed. To illustrate the efficiency of employing a series of solid-supported catalysts in a microreactor, the authors performed an analogous batch reaction and found that a reaction time of 24 h was required in order to obtain comparable conversions (cf. ~1 min residence times within the microreactor).

Having demonstrated the ability to incorporate multiple heterogeneous catalysts into an EOF-based flow reactor, the authors evaluated an increase in packed-bed size as a means of increasing the throughput of the flow reactor from mg h^{-1} to g h^{-1}. As the diffusion distance between the reactant solution and the solid-supported catalysts remains the same, the reaction efficiency attained in the capillary reactor was maintained and the increased flow rate employed afforded a higher throughput. In practice, this was achieved by increasing the reaction channel to 3 mm (i.d.), packing with A-15 **126** (36 mg) and silica-supported piperazine **113** (50 mg) and employing a flow rate of 54.9 μl min^{-1} afforded a throughput of 0.93 g h^{-1} in analytical purity (Wiles et al., 2007d).

In another example employing multiple supported catalysts and reagents, Smith et al. (2007a) presented a modular flow reactor in which 14 1,4-disubstituted-1,2,3-triazoles were synthesized. Coupling an immobilized copper(I) iodide species **148** with two scavenger modules (immobilized thiourea **149** and phosphane resin **150**), the authors reported the [3 + 2] cycloaddition of an array of azides and terminal acetylene (30 μl min^{-1}) to afford the desired 1,4-disubstituted 1,2,3-triazoles (Scheme 40) in moderate to excellent yield (70–93%).

In analytical mode, the reactor was optimized to afford between 20 and 200 mg of product; however in the case of propargylic alcohol

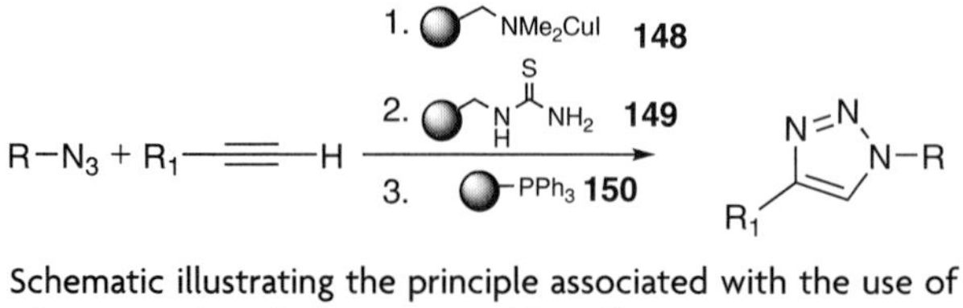

Scheme 40 Schematic illustrating the principle associated with the use of multiple solid-supported reagents in a single pressure-driven flow reactor.

($R^1 = CH_2OH$) and benzyl azide ($R = CH_2PH$), the reactor was operated continuously for 3 h to afford 1.50 g of the desired product in 85% yield and 95% purity. Additional examples reported by the group included amides and guanidines whereby purification was facilitated by the use of tagged phosphane reagents (Smith et al., 2007b).

Wiles and Watts (2007b) developed a microreactor to enable the parallel screening of catalysts and supported reagents, using the synthesis of tetrahydropyranyl ethers as a model reaction (Table 22). Employing a borosilicate glass microreactor containing four identical reaction channels [280 μm (wide) × 90 μm (deep) × 2.0 cm (long)], the authors evaluated a series of polymer-and silica-supported Lewis acid catalysts for their efficiency toward the protection of benzyl alcohol **35** as 2-(benzyloxy) tetrahydropyran **151**. In each case, 1 mg of catalyst was employed and as Table 22 illustrates, excellent conversions were obtained for all catalysts; however, the higher loading of the silica-supported sulfonic acid

Table 22 Summary of the results obtained for the synthesis of 2-(benzyloxy)tetrahydropyran **151**, employing an array of solid-supported catalysts

Channel no.	Supported Lewis acid	Loading (mmol g^{-1})	Flow rate (μl min^{-1})	Conversion (%)
1		3.50	1.10	100.0 (0.0)[a]
2	**135**	0.60	1.60	99.9 (3.5×10^{-4})[a]
3	**152**	4.20	1.80	100.0 (0.0)[a]
4	(SO$_3$)$_3$ Yb	2.00	1.50	99.9 (2.6×10^{-4})[a]

[a]The numbers in parentheses represent the % RSD, where $n = 15$.

derivative **152** enabled the reactions to be performed at a high throughput. Having identified the most active catalyst, the reactor was subsequently used to protect a further 14 alcohols whereby isolated yields ≥ 99.4% were obtained; in addition, employing MeOH as the reaction solvent enabled facile deprotection of the THP ethers. Using this approach, catalytic turnover numbers in excess of 2,760 were obtained with the catalyst showing no sign of degradation.

2.4.5 Biocatalysis

In addition to the wide range of chemically catalyzed reactions conducted in microreactors, the past 2 years has seen a large amount of interest in the use of biocatalysts within such systems (Honda et al., 2006). Of these, one of the most noteworthy examples is the fused silica microreactor reported by Belder et al. (2006), who fabricated an integrated system capable of synthesis, separation and detection. Using the enantioselective synthesis of 3-phenoxypropane-1,2-diol **153** as a model reaction, the authors evaluated the hydrolysis of 2-phenoxymethyloxirane **154** using an epoxide hydrolase enzyme (Scheme 41). The next step of the process involved the electrophoretic separation of the resulting reaction mixture, which was achieved in 90 s, followed by fluorescence detection of the two enantiomers using a deep-UV laser (Nd:YAG, 266 nm). Using this technique enabled three mutants of the epoxide hydrolase *Aspergillus niger* affording conversions in the range of 22–43% and *ee* values of 49–95%.

In 2006, Wang et al. (2006) reported the fabrication and evaluation of a polydimethylsiloxane (PDMS) microreactor capable of performing 32 enzyme-catalyzed click reactions in parallel (Table 22). The reactor consisted of several components including a nanoliter rotary mixer, a chaotic mixer, and a microfluidic multiplexer, which enabled discrete aliquots of reactants (57 s reactant^{-1}) to be introduced into the reaction channel and allowed multiple reactions to be performed in parallel. Using this approach, the authors performed 10 *in situ* click reactions between acetylene **155** and 10 different azides in the presence of (1) bovine carbonic anhydrase II (bCAII) **156**, (2) bCAII **156** and an inhibitor, (3) in the absence of bCAII **156**, and (4) two blank solutions containing only bCAII **156** and PBS solution. Using this approach, the reactions were performed in two batches and employing DMSO/EtOH (1:4) as the reaction solvent, the

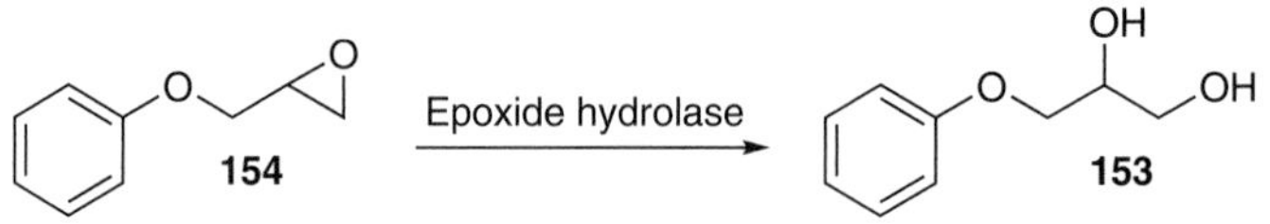

Scheme 41 Model reaction used to demonstrate the biocatalytic hydrolysis of 2-phenoxymethyloxirane **154**.

reaction mixtures were subsequently heated at 37 °C for 40 h prior to analysis by LC–MS. In addition to the speed of processing, the microreactor also enabled the quantities of reactants to be reduced with a typical reaction consuming 4 µl of reaction mixture (19 µg bCAII **156**, 2.4 nmol acetylene **155**, 3.6 nmol azide) [cf. 100 µl (94 µg bCAII **156**, 6.0 nmol acetylene **155**, 40.0 nmol azide) in a batch protocol]; a 2–12-fold reduction depending on the reactant. The authors were also pleased to confirm that the same results were obtained in the microreactor as within a 96-well plate (batch reaction), reporting the same nine compounds as hits out of 20 click compounds; a selection of which are illustrated in Table 23.

In order to increase the efficiency of biocatalytic transformations conducted under continuous flow conditions, Honda et al. (2006, 2007) reported an integrated microfluidic system, consisting of an immobilized enzymatic microreactor and an in-line liquid–liquid extraction device, capable of achieving the optical resolution of racemic amino acids under continuous flow whilst enabling efficient recycle of the enzyme. As Scheme 42 illustrates, the first step of the optical resolution was an enzyme-catalyzed enantioselective hydrolysis of a racemic mixture of acetyl-D,L-phenylalanine to afford L-phenylalanine **157** (99.2–99.9% *ee*) and unreacted acetyl-D-phenylalanine **158**. Acidification of the reaction products, prior to the addition of EtOAc, enabled efficient continuous extraction of L-phenylalanine **157** into the aqueous stream, whilst acetyl-D-phenylalanine **158** remained in the organic fraction (84–92% efficiency). Employing the optimal reaction conditions of 0.5 µl min^{-1} for the enzymatic reaction and 2.0 µl min^{-1} for the liquid–liquid extraction, the authors were able to resolve 240 nmol h^{-1} of the racemate.

Employing a multichannel PDMS microreactor [350 µm (wide) × 250 µm (deep) × 6.4 mm (long)], in which the thermophilic enzyme β-glycosidase was immobilized, Thomsen et al. (2007) evaluated the hydrolysis of 2-nitrophenyl-β-D-galactopyranoside. Heating the reactor to 80 °C, the authors were able to continuously hydrolyze 2-nitrophenyl-β-D-galactopyranoside and monitored the reaction efficiency via generation of 2-nitrophenol **97**.

Again using the principle of PASSflow, Drager et al. (2007) recently reported the development of a polymeric matrix that was capable of performing automatic purification and immobilization of His$_6$-tagged proteins, followed by their use as highly active biocatalysts. Using the polymeric support illustrated in Figure 1, the authors conducted the immobilization of two enzymes, benzaldehyde lyase (BAL) and *Bacillus subtilis* (BsubpNBE), and evaluated the resulting biocatalysts toward the benzoin reaction (Scheme 43) and the regioselective hydrolysis of esters.

Initial investigations into the efficacy of the immobilization strategy were conducted using the benzoin reaction whereby employing benzaldehyde **116** (5.5 × 10^{-2} M) in phosphate buffer (pH 7, containing 10%

Table 23 Summary of the 20 click reactions conducted within a PDMS microreactor, highlighting those which were identified to be a hit

Azide	Result	Azide	Result
	Hit		Hit
	Hit		Hit
	No hit		No hit

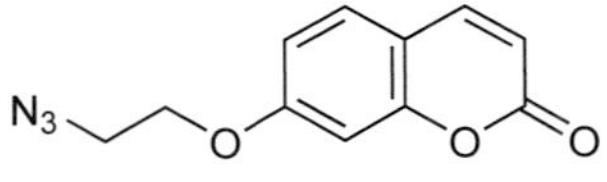	Hit	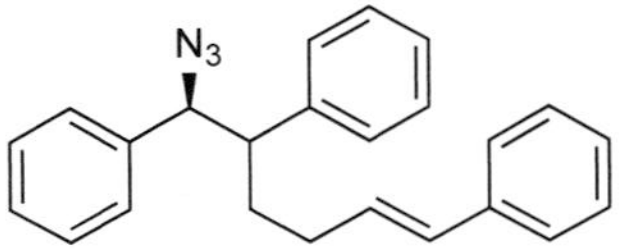	No hit
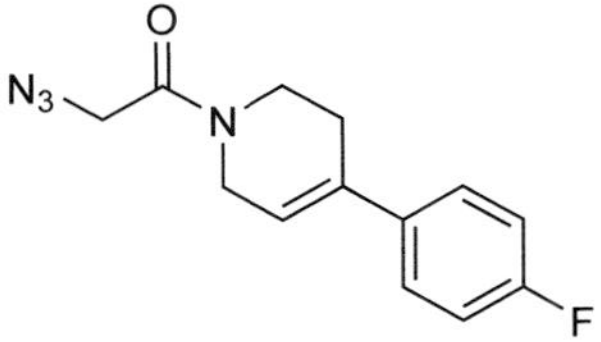	Hit	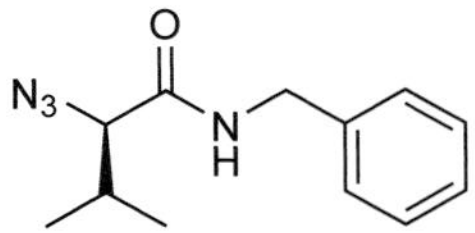	Hit
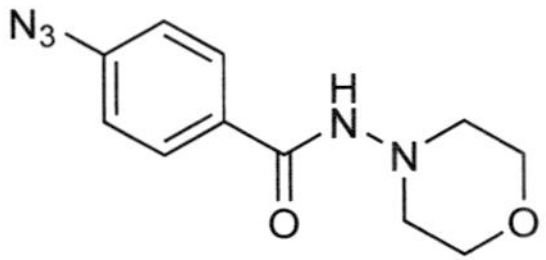	No hit	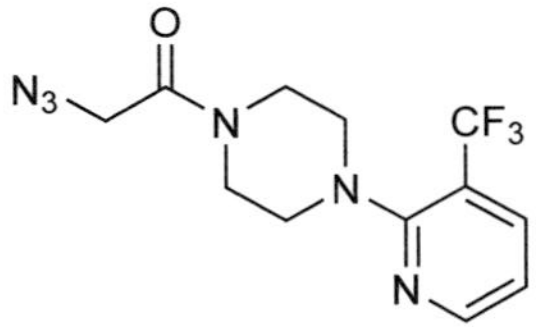	No hit
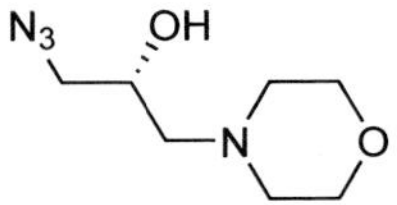	No hit	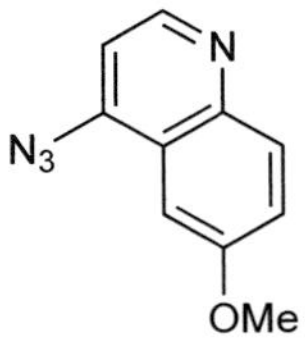	No hit

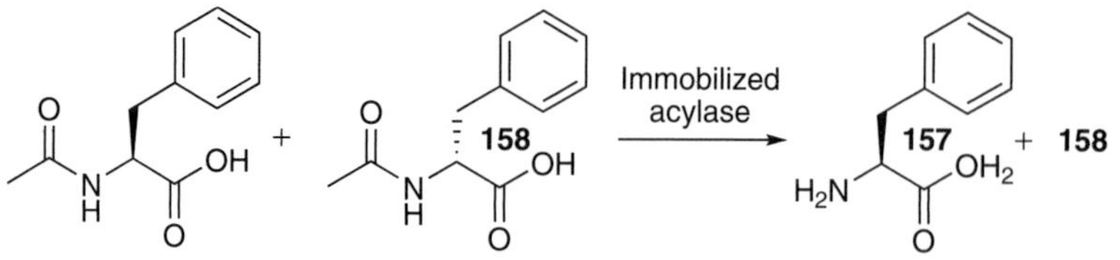

Scheme 42 Continuous flow optical resolution of acetyl-D,L-phenylalanine using an immobilized aminoacylase enzyme.

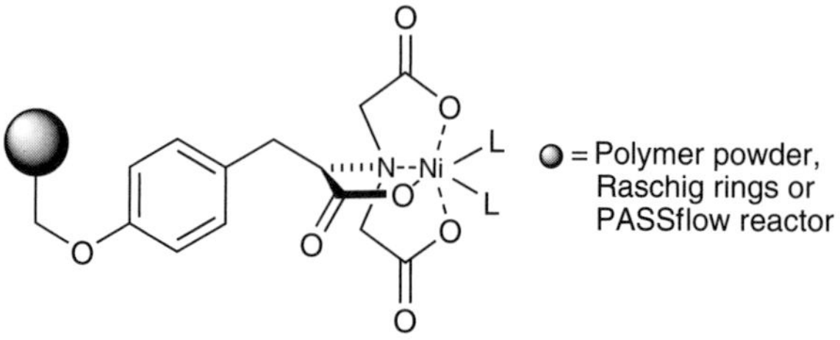

Figure 1 Illustration of the tyrosine-based matrix used for the immobilization of His$_6$-tagged proteins.

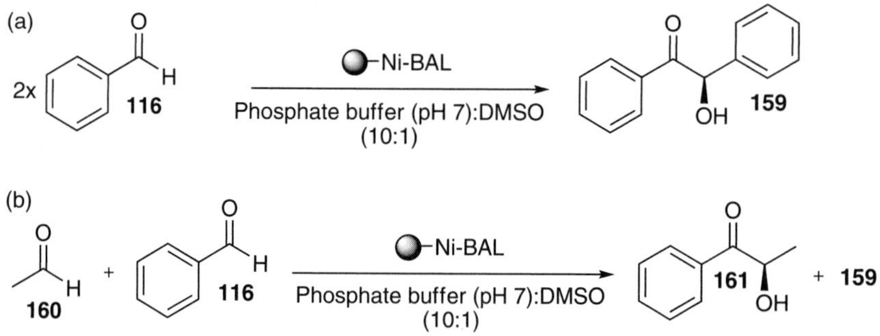

Scheme 43 An example of the (a) benzoin reaction and (b) cross-benzoin reaction conducted using immobilized His$_6$-tag BAL.

DMSO), a reaction temperature of 37 °C, and a residence time of 90 min, achieved by recirculating the reaction mixture through the PASSflow reactor at a flow rate of 1.0 ml min^{-1}, the authors were able to attain 99.5% conversion of **116** to (R)-benzoin **159**, determined by off-line GC analysis. Increasing the reactant concentration from 5.5×10^{-2} to 0.2 M, resulted in a reduction in benzoin **159** production of 5.9%, with longer reaction times required to attain high conversions with further increases in reactant concentration (typically 9 h for 1.0 M **116**). The authors subsequently demonstrated the cross-benzoin reaction between acetaldehyde

160 (2.3×10^{-1} M) and benzaldehyde **116** (4.6×10^{-2} M) whereby the target compound (R)-2-hydroxy-1-phenylpropan-1-one **161** was obtained in 92% isolated yield; (R)-benzoin **159** was obtained as a minor by-product in 8% yield.

Having confirmed that the immobilization strategy was robust enough to be used in conjunction with a continuous flow reactor, the authors evaluated the immobilization of a second His$_6$-tag enzyme, BsubpNBE, and employed the synthesis of 6-O-acetyl-D-glucal **162** (Scheme 44) as a model reaction.

In comparison to the previous example, the ester hydrolysis was significantly slower requiring a reaction time of 60 h to consume tri-O-acetyl-D-glucal **163** and afford the target compound **161** in 80% yield, with D-glycal **164** as a minor by-product (12.0%).

Along with problems associated with enzyme recovery and reuse, difficulties associated with the efficient recycle of cofactors have also contributed to the limited industrial uptake of biocatalysis. With this in mind, Yoon et al. (2005) reported a PDMS microfluidic system [channel dimensions = 250 µm (wide) × 3 cm (long)] capable of electrochemically regenerating nicotinamide cofactors and thus reducing the costs associated with the use of enzymes that require the presence of a cofactor. The authors found that by employing multistream laminar flow, comprising of a buffer stream and a reagent stream, regeneration of the cofactor could be achieved at the surface of a gold electrode. Using the conversion of achiral pyruvate **165** to L-lactate **166** in the presence of an enzyme (Scheme 45), the authors were able to demonstrate efficient enzyme/cofactor regeneration, equivalent to a turnover number of $75.6\,h^{-1}$.

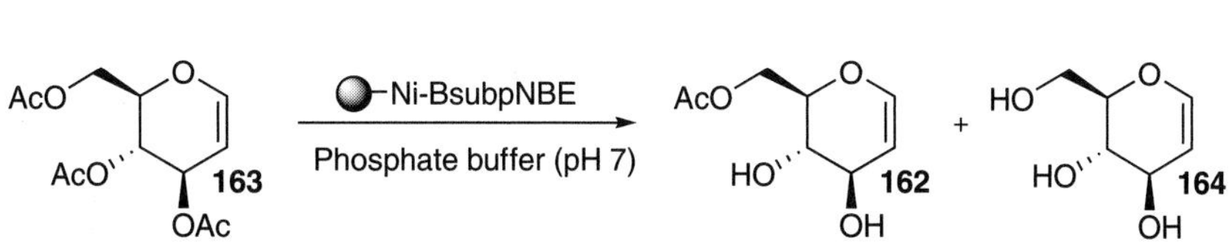

Scheme 44 The continuous flow ester hydrolysis achieved using a flow reactor containing immobilized His$_6$-tag BsubpNBE.

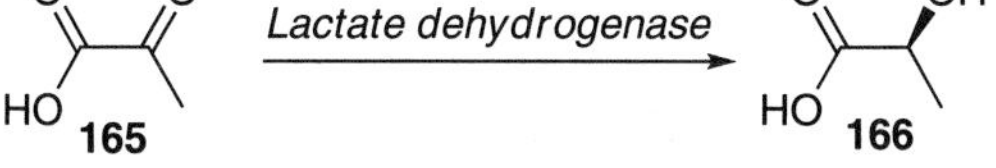

Scheme 45 Enzymatic synthesis of L-lactate **166**, employing electrochemical cofactor regeneration.

Csajagi et al. (2008) recently demonstrated the enantioselective acylation of racemic alcohols in a continuous flow bioreactor, using *Candida antarctica* lipase B (CaLB) **167**. Employing a packed-bed reactor, containing 0.40 g of enzyme **167**, and pumping a solution of *rac*-phenyl-1-ethanol **119** (10 mg ml^{-1}) in hexane:THF:vinyl acetate **168** (2:1:1) at a flow rate of 100 μl min^{-1} (at 25 °C), the authors found the reactor reached steady state after 30 min of operation. Under the aforementioned conditions, the (*R*)-acetate **169** was obtained in 50% conversion and 99.2% *ee* and residual (*S*)-alcohol **170** in 98.9% *ee* with a residence time of 8.2 min; analogous results were obtained in batch, however, required 24 h to afford 50% conversion of **119–169**. Having devised a rapid technique for the evaluation of immobilized biocatalysts, the authors compared a series of lipase enzymes for the model reaction illustrated in Scheme 46, whereby CaLB **167**, lipase *Pseudomonas cepacia* IM, and Amano lipase AK were found to afford the highest throughputs of 10.2, 10.2, and 10.6 μmol min^{-1} g^{-1}, respectively.

To demonstrate the synthetic application of this methodology, the authors subsequently demonstrated its use for the preparative kinetic resolutions of a series of 2° alcohols, Table 24, whereby 20 ml solutions of each racemic alcohol were passed through the bioreactor (3.3 h) and found to afford analogous results to those obtained during the initial optimization experiments. The authors successfully demonstrated the use of immobilized and lyophilized enzymes within a continuous flow reactor, presenting a synthetically viable approach to the kinetic resolution of racemic alcohols.

Due to their cost, instability and limited longevity, enzymes are not widely employed in production scale syntheses; however, through their incorporation into flow reactors, biocatalysts can be readily employed in the synthesis of high value products.

2.5 The use of solid-supported reagents in noncatalytic flow processes

From the examples above it can be seen that the incorporation of solid-supported catalysts into flow reactors can afford extremely efficient processes; however, recently numerous authors have reported the use of

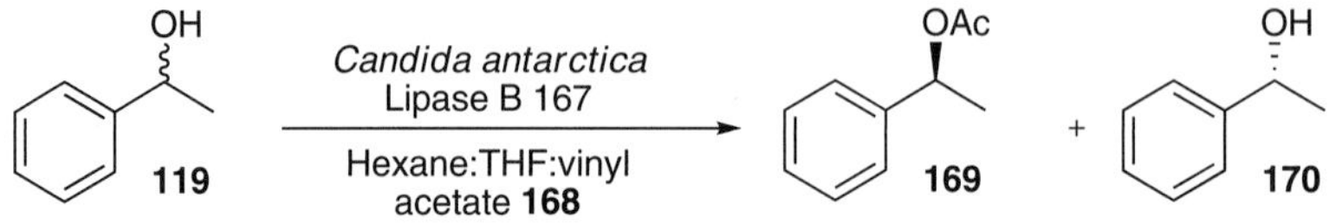

Scheme 46   Schematic illustrating the biochemical acylation of *rac*-phenyl-1-ethanol **119**.

Table 24 Examples of the preparative-scale kinetic resolutions conducted under flow

Compound	Yield (%)[a]	ee[b]	$[\alpha]_D^{25c}$	E[d]
OAc (R)-**169**	40	98.5	−62.8	
OH (S)-**170**	48	99.1	+125.3	>>>200
OAc (R)⁻	26	77.4	+2.0	
OH (S)⁻	41	99.0	+7.1	>200
OAc (R)⁻	34	56.4	+4.9	
OH (S)⁻	41	85.1	−23.3	22

[a]Products isolated from the reactor output stream.
[b]Determined by enantioselective GC.
[c]Specific rotations (~1.0, CHCl₃).
[d]Due to sensitivity to experimental errors, enantiomer selectivity values calculated in the range 25–500 are reported as >200 and those above 500 as >>>200.

solid-supported reagents and scavengers, which have a finite lifetime in the absence of regeneration techniques, under flow conditions and illustrated a series of novel techniques that are useful to the modern day synthetic chemist.

Over the past decade, numerous authors have reported the continuous flow synthesis of peptides via serial deprotection and coupling steps (Flogel et al., 2006; Watts et al., 2002), to date, there have been no reports of continuous flow protections, followed by reaction of an unprotected moiety and selective deprotection. Owing to the complex nature of protecting group chemistry, it was the aim of a recent study, conducted by Wild et al. (2009), to develop a noncovalent protecting group strategy which would enable the facile N-protection of bifunctional compounds and subsequent reaction of the unprotected functionality to afford the target molecule in excellent yield and selectivity. With this aim in mind, the authors investigated the use of crown ethers as noncovalent protecting groups as they have been shown to efficiently sequester ammonium ions, forming a stable adduct via hydrogen bonding. Using this principle, several authors have reported the use of crown ethers as N-protecting groups in solution; however, widespread application of the methodology has been precluded by problems associated with product isolation and reuse of the crown ether. To address this, Wild et al. (2009) prepared a solid-supported 18-crown-6 ether derivative and subsequently demonstrated its use as a novel N-protecting group within a packed-bed flow reactor. To illustrate the selectivity of the technique, the O-acetylation of tyramine **171** was conducted using acetyl chloride **19**, in the absence of a protecting group, which resulted in a complex mixture consisting of the desired tyramine acetate **172** (23%), tyramine N-acetate **173** (12%), tyramine diacetate **174** (20%), and residual starting material (45%) (Scheme 47), when conducted in a batch reaction.

In comparison, employing the reaction strategy depicted in Scheme 48 whereby the trifluoroacetic acid (TFA) salt of tyramine **175** was N-protected by the immobilized crown ether [step (a)], O-acetylated with acetic anhydride **37** in the presence of an organic base **14** [step (c)], and then simultaneously deprotected and the crown ether regenerated

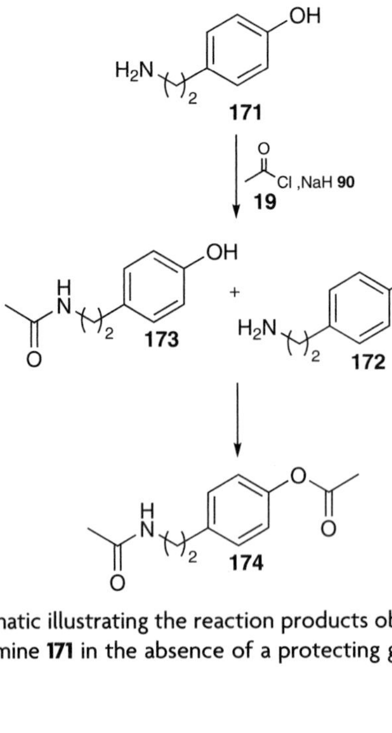

Scheme 47   Schematic illustrating the reaction products obtained when conducting the acetylation of tyramine **171** in the absence of a protecting group.

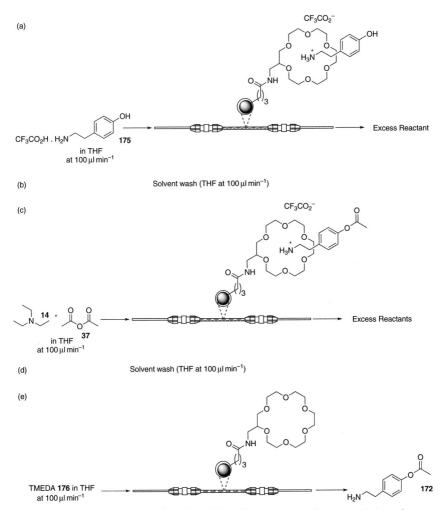

Scheme 48 Reaction protocol employed for the continuous flow acetylation of tyramine, using an immobilized 18-crown-6 ether derivative (0.15 g, 0.16 mmol g^{-1}).

using a solution of $N,N,N'N'$-tetramethylethylenediamine **176** [step (e)], the target compound tyramine acetate **172** was obtained in quantitative yield (2.4×10^{-2} mmol reaction^{-1}) and selectivity. The authors also investigated the generality of the protecting group strategy finding that HCl, TFA, and p-TSA salts could be complexed readily and several bifunctional compounds were also reacted to afford the O-product in all cases. It is acknowledged that automation of the methodology would increase the efficiency of the process, making it a practical alternative to conventional solution-phase protection/deprotection chemistry.

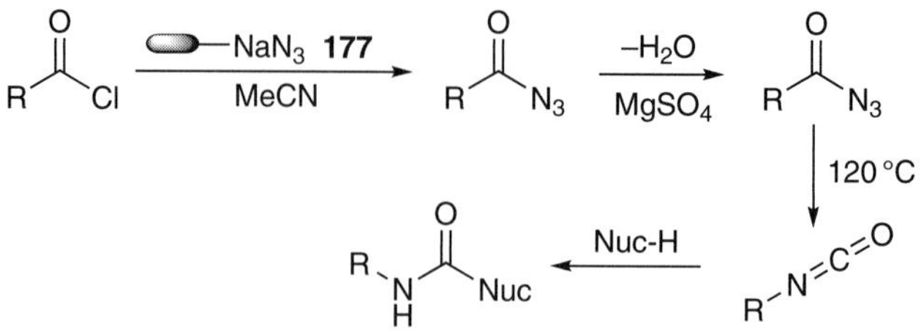

Scheme 49 Generalization of the polymer-assisted Curtius rearrangement conducted under continuous flow.

One of the more recent solid-supported reagents utilized by Ley and co-workers is an azide monolith, used to synthesize a series of azides from their respective acyl halide, as illustrated in Scheme 49 (Baumann et al., 2008a). The authors report the fabrication of the azide monolith **177** $(2.00 \, mmol \, g^{-1})$ in a glass Omnifit reactor [15 mm (i.d.) × 10 cm (long)] via thermal polymerization and quantified the columns total void volume (6.4 ml) using mercury intrusion analysis.

To synthesize a series of organic azides, the authors pumped a solution of the acyl chloride (1.0 M in MeCN) through the azide monolith **177**, at room temperature $(0.5 \, ml \, min^{-1})$, to afford a residence time of 13 min. In the case of the substrate 3-bromobenzoyl chloride, removal of the reaction solvent *in vacuo* afforded the respective azide, 3-bromobenzoyl azide, in quantitative yield and excellent purity. Having demonstrated the successful conversion of acyl chlorides to acyl halides, the next step of the investigation was to decompose the acyl azide under flow, *in situ*, to afford the respective isocyanate.

As Scheme 49 illustrates, this was achieved by firstly passing the acyl azide solution through a drying agent to remove any residual moisture and then superheating the reaction mixture in a convection flow coil to 120 °C. The isocyanate was then collected in a sealed vial, containing a nucleophile (1.0–4.0 eq.), and the reaction mixture heated to 100 °C, for 10 min, in a microwave to afford the reaction products summarized in Table 25. In both cases, the azide monolith **177** was used as a stoichiometric reagent and as such required regeneration after 1–10 mmol of organic azide had been prepared. Monolith regeneration proved facile and was achieved by simply pumping a solution of sodium azide through the depleted column to afford the original loading. The authors also reported the Curtius rearrangement starting from carboxylic acids and demonstrated the simultaneous trapping of the isocyanate intermediate with an array of nucleophiles to afford the target compounds in 75–95% yield and >90% purity (Baumann et al., 2008b).

Table 25 Summary of the products generated using azide monoliths $(2.0\,\text{mmol}\,\text{g}^{-1})$ under continuous flow

Product	Yield (%)
	84
	73
	90
	64
	76
	78

The selective oxidation of primary alcohols is notoriously difficult to achieve selectively as further oxidation of the product to the respective carboxylic acid occurs rapidly, as depicted in Scheme 50. Wiles et al. (2006) therefore proposed that it should be possible to isolate either the aldehyde or the carboxylic acid, depending on the reaction times employed. To demonstrate this, the authors constructed a packed-bed reactor [3 mm (i.d.) × 5.0 cm (long)] containing silica-supported Jones' reagent (0.15 g, 0.15 mmol), a Cr(VI)-based oxidant, and by exploiting the high surface to volume ratio obtained within continuous flow reactors, the

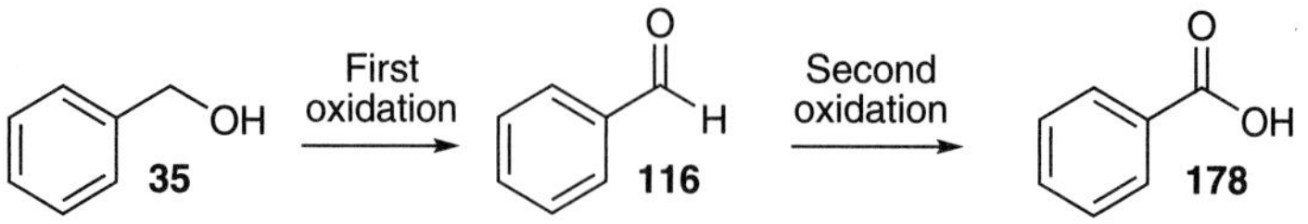

Scheme 50 Illustration of the overoxidation of benzyl alcohol **35** to benzoic acid **178**, observed in a stirred batch reactor.

authors were able to selectively oxidize an array of aromatic primary alcohols to either the aldehyde or carboxylic acid, depending on the reactant residence time within the packed bed.

Under pressure-driven flow, a reactant stream of the aromatic alcohol under investigation (1.0×10^{-2} M) in DCM was pumped through the packed bed and the effect of residence time evaluated by varying the flow rate; reaction products were analyzed by GC–MS and the conversion to the aldehyde and carboxylic acid was quantified. Conducting the first reactions at a flow rate of $300 \, \mu l \, min^{-1}$ (21 s) afforded consumption of the alcohol **35** to afford a mixture of the aldehyde **116** and carboxylic acid **178**, reducing the flow rate to $50 \, \mu l \, min^{-1}$ (126 s) converted all available aldehyde **116** to benzoic acid **178**. Consequently, in order to selectively synthesize benzaldehyde **116**, the flow rate was increased to $650 \, \mu l \, min^{-1}$, affording a reaction time of 9.7 s, and the aldehyde **116** isolated in quantitative yield. Having successfully identified a means of the chemoselective oxidation of primary alcohols, a series of substituted alcohols were investigated, demonstrating no substrate dependency with respect to aromatic alcohols (Table 26). In addition to affording a chemoselective technique, the use of a solid-supported oxidant was advantageous as no metal contaminated waste streams were generated, as is the case with homogeneous Cr(VI)-based oxidants.

Unfortunately, the technique was not suitable for the oxidation of aliphatic aldehydes as the increase in polarity of the reaction stream (cf. aromatic alcohols) led to leaching of the Jones' reagent from the solid support. The methodology did however enable the facile oxidation of secondary alcohols to their respective ketone, as no competing overoxidation products could be formed, the reactions were conducted at a flow rate of $650 \, \mu l \, min^{-1}$ (residence time $= 9.7$ s) to again afford the respective ketones in excellent yield (99.0–100.0%) and quantitative purity. Although the oxidant employed was not catalytic, the solid-supported reagent was self-indicating and turned from orange to green upon exhaustion.

2.6 Photochemical reactions

Along with improving those synthetic transformations that are routinely employed in the development and identification of lead compounds, there are many interesting synthetic procedures that are investigated within

Table 26 Summary of the results obtained for the chemoselective oxidation of primary alcohols

Alcohol	Flow rate (µl min^{-1})	Product distribution (%)	
		Aldehyde	Carboxylic acid
Benzyl alcohol **35**	650	100 (99.1)[a]	0
	50	0	100 (99.6)
3,5-Dimethoxybenzyl alcohol	650	100 (99.5)	0
	50	0	100 (99.3)
4-Bromobenzyl alcohol	650	100 (99.0)	0
	50	0	100 (98.3)
4-Chlorobenzyl alcohol	650	100 (99.3)	0
	50	0	100 (99.4)
4-Cyanobenzyl alcohol	650	100 (98.5)	0
	50	0	100 (99.0)
Methyl-4-formylbenzyl alcohol	650	100 (99.2)	0
	50	0	100 (99.6)
4-Methylbenzyl alcohol	650	100 (99.2)	0
	50	0	100 (95.6)
4-Benzyloxybenzyl alcohol	650	100 (99.5)	0
	50	0	100 (99.8)
4-Aminobenzyl alcohol	650	100 (100)	0
	50	0	100 (99.8)
4-Dimethylaminobenzyl alcohol	650	100 (99.3)	0
	50	0	100 (99.6)
Biphen-4-yl methanol	650	100 (99.7)	0
	50	0	100 (99.5)
(5-Nitrothiophen-2-yl)-methanol	650	100 (99.8)	0
	50	0	100 (99.7)
2-Benzyloxybenzyl alcohol	650	100 (99.7)	0
	50	0	100 (99.8)
2-Naphthalen-2-yl methanol	650	100 (99.9)	0
	50	0	100 (99.9)
4-Acetylbenzylalcohol	650	100 (99.8)	0
	50	0	100 (99.8)

[a]The number in parentheses represents the isolated yield (%).

research laboratories which fail to be adopted when it comes to target manufacture. One such technique is photochemistry, which despite its ability to efficiently generate reactive intermediates many contributing factors have precluded the use of photochemistry on a large scale, such as difficulties associated with scaling light sources. As such, photochemical syntheses represent one of the most interesting applications of this emerging technology as MRT has the potential to enable the many synthetically useful photochemical transformations reported by academic researchers to be used for the industrial-scale synthesis of fine chemicals and pharmaceutical agents. The following discussion therefore highlights the current problems associated with photochemistry, and subsequently describes how the microreaction technology has the potential to resolve the most fundamental of these issues.

2.6.1 Homogeneous photochemistry

Utilizing a commercially available microreactor, fabricated from FOR-TURAN® glass, Fukuyama et al. (2004) evaluated a series of [2 + 2] cycloadditions as a means of reducing the reaction times conventionally associated with the synthetic transformation (Table 27). Using a high-pressure mercury lamp (300 W), the reaction of cyclohex-2-eneone **179** with vinyl acetate **168** (Scheme 51), to afford the cycloadduct **180**, was used to compare photochemical efficiency within the microreactor [1,000 μm (wide) × 500 μm (deep)] and a conventional batch reactor (10 ml).

In line with the literature, irradiating the batch reactor for 2 h afforded only yielded 8% **180**; however, employing a residence time of 2 h within the microreactor, achieved by pumping the reaction mixture through the microchannel at a flow rate of 8.3 μl min^{-1}, the authors obtained adduct **180** in 88% yield. The enhanced irradiation efficiency obtained within the flow reactor therefore enabled a dramatic increase in reaction yield, coupled with a reduction in the overall reaction time required. With this in mind, the authors investigated the generality of the technique and as Table 26 illustrates, moderate to good yields were obtained for a range of substituted cyclohex-2-enones and vinylic compounds.

Another example of this was reported by Hook et al. (2005), whereby a series of photocycloadditions were performed using a standard, water-cooled immersion well, wrapped in fluorinated ethylene propylene (FEP) tubing (channel dimensions = 700 μm). In the first instance, the authors demonstrated the synthesis of 6-butyl-3-azabicyclo[3.2.0]hept-6-ene-2,4-dione **181** (Scheme 52a), which was achieved by pumping a solution of maleimide **182** (0.10 M) and 1-hexyne **183** (0.15 M), in MeCN, through the reactor at 2.0 ml min^{-1}; affording the target compound **181** in 95% yield with a throughput of 2.1 g h^{-1}. Additional photochemical transformations performed by the group included the synthesis of 7,8-dimethyl-1,2,3,9a-tetrahydropyrrolo[1,2-a]azepine-6,9-dione **184** (Scheme 52b) which was obtained in 80.0% yield at an impressive throughput of 7.4 g h^{-1}.

Mukae et al. (2007) also compared a Pyrex batch reaction vessel (8 mm) with several Pyrex microreactors [channel dimensions = 100 μm (wide) × 40 μm (deep) 1.20 cm (long)] for the photocycloaddition of

Table 27 Photochemical [2 + 2] cycloaddition conducted in a FORTURAN® glass microreactor

Substituted cyclohex-2-enone	Vinylic substrate	Cycloadduct	Irradiation time (h)	Yield (%)
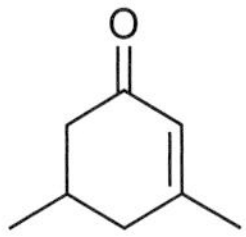	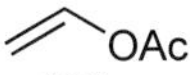 **168**	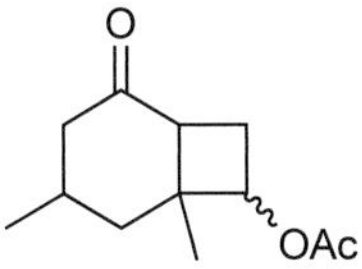	2	70
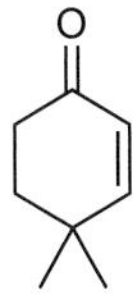	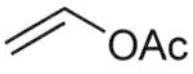 **168**	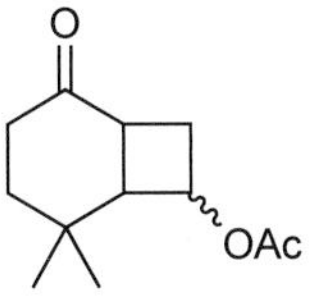	3.2	62
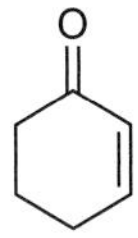 **179**	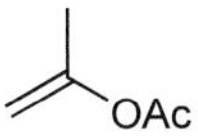	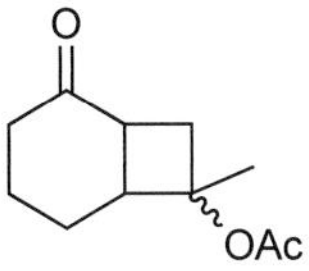	3.2	64
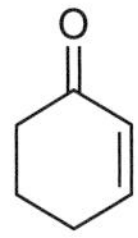 **179**	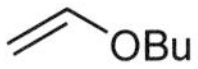	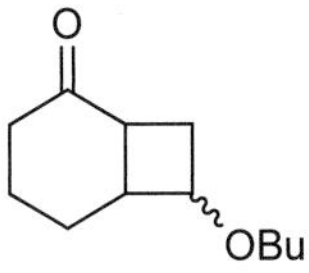	3.2	67

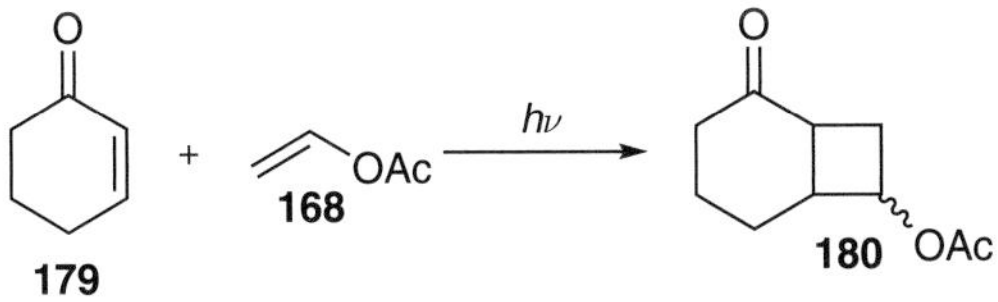

Scheme 51 Photochemical [2 + 2] cycloaddition of cyclohex-2-eneone **179** with vinyl acetate **168**.

2-(2-alkenyloxymethyl)-naphthalene-1-carbonitriles **186**, as depicted in Scheme 53. Using a xenon lamp (500 W) as the light source, the effect of alkene substitution, solvent, residence time, and microchannel dimensions on the cycloaddition was evaluated.

Under standard batch conditions, the target products were obtained in 74% yield with an irradiation time of 3 h, compared with analogous yields in the microreactor with an irradiation time of only 1 min. In addition to

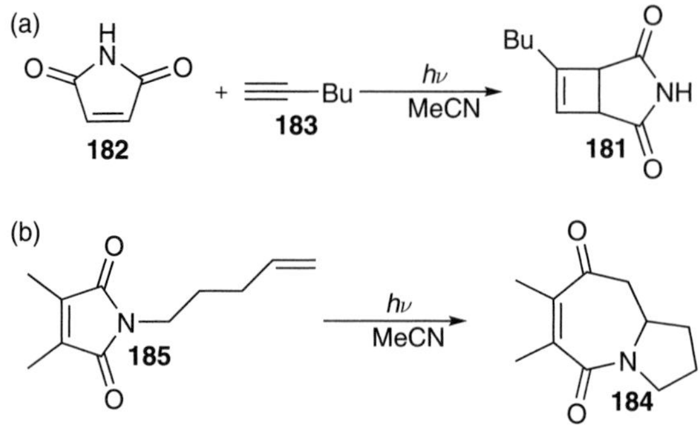

Scheme 52 (a) [2 + 2] Photocycloaddition of maleimide **182** and 1-hexyne **183** to afford **181** and (b) intramolecular [5 + 2] photocycloaddition of 3,4-dimethyl-1-pent-4-enylpyrrole-2,5-dione **185**.

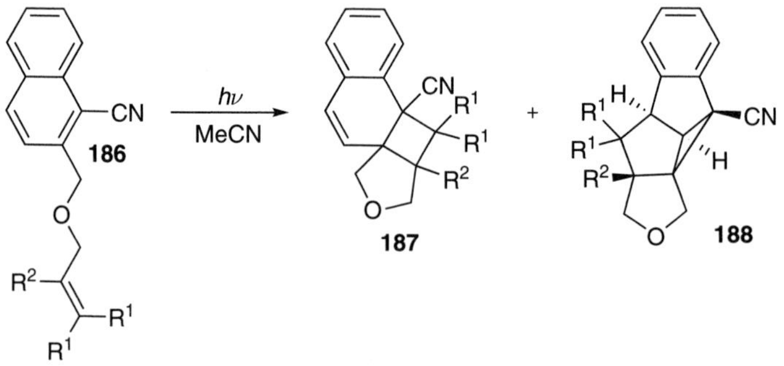

Scheme 53 Photochemical synthesis of a series of 2-(2-alkenyloxymethyl)-naphthalene-1-carbonitriles.

the obvious increase in photon efficiency obtained, the authors also noted an increase in reaction regioselectivity when conducting the reaction under flow conditions an observation which is attributed to the ability to remove the initial products **187** from the reactor prior to the second reaction, an undesirable photocycloreversion, which affords products of the type **188**. In addition, the authors also demonstrated the ability to boost the throughput of the reactor by simply increasing the width of the reaction channel (100–2,500 μm) whilst maintaining the original channel depth, illustrating the ease with which such reactions can be scaled. Using this approach a 25-fold increase in reactor throughput was obtained, with no observable decrease in reaction selectivity.

With numerous researchers investigating the advantages associated with thermally or biocatalytically controlled asymmetric syntheses, some of which have been performed in continuous flow reactors, few have considered the prospects of photochemical asymmetric synthesis, an idea

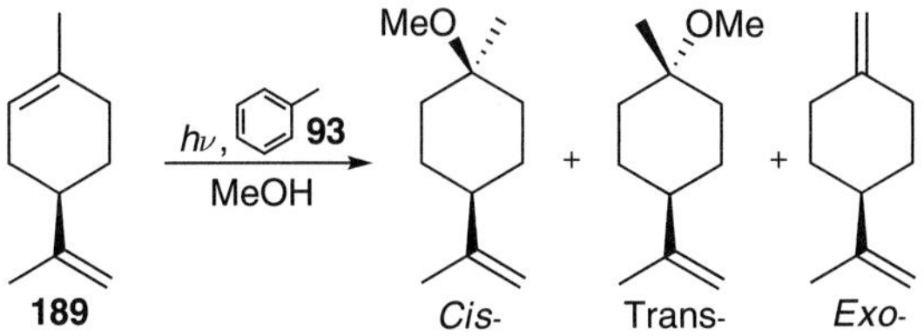

Scheme 54 Asymmetric addition of MeOH to (R)-(+)-(Z)-limonene (**189**) performed in a quartz microreactor.

recently communicated by Sakeda et al. (2007). Using the asymmetric photochemical addition of methanol to (R)-(+)-(Z)-limonene (**189**) as a model reaction (Scheme 54), the authors compared three quartz microreactors, with a standard laboratory cell as a means of highlighting the synthetic potential of this approach.

Employing a low-pressure Hg lamp ($\lambda = 254$ nm) the authors evaluated the effect of channel geometry on photon efficiency and diastereomeric excess (*de*) (see Table 28) by pumping a methanolic solution of (R)-(+)-(Z)-limonene (**189**) (25 mM) and toluene **93** (10 mM) through the various quartz reactors. Initial investigations were conducted using a microreactor with channel dimensions of 500 μm (wide) × 300 μm (deep) and focused on determining the effect of irradiation time on the conversion of (R)-(+)-(Z)-limonene (**189**). Using this approach, a linear relationship between irradiation time and conversion was observed; however, the authors did observe a decrease in *de* with prolonged exposure times. Based on this observation, a series of microchannel geometries were subsequently evaluated and it was found that shallow channels, typically <40 μm, afforded increased photon efficiency; which is attributed to illumination homogeneity. Furthermore, a slight increase in *de* was observed in all flow experiments when compared to the batch cell, a feature that was explained by the suppression of side reactions within the continuous flow reactors.

Table 28 Summary of the results obtained for the asymmetric addition of MeOH to (R)-(+)-(Z)-limonene (**189**) in batch and a microreactor with an irradiation time of 36 s

Reactor	Dimensions	Photon efficiency	*de* (%)
Batch	100 μm × 3 mm	0.06	28.7
Micro	500 μm × 300 μm	0.11	30.6
Micro	400 μm × 40 μm	0.27	29.4
Micro	200 μm × 20 μm	0.29	30.0

2.6.2 Heterogeneous photochemistry

The homogeneous photochemical reactions described thus far serve to illustrate the effectiveness achieved as a result of combining photochemistry and microreactors; however, organic photochemistry is not limited to homogeneous reactions, with many synthetically useful transformations conducted using catalytic processes. Miniaturized catalytic photochemistry initially focusing on the photodegradation of substrates (Gorges et al., 2004; Li et al., 2003); however, more recently researchers have reported the use of such systems for a series of common organic reactions including reductions, alkylations, and cyclizations, a selection of which are discussed below.

An early example of a photochemical cyclization was reported by Takei et al. (2005) who demonstrated the synthesis of L-pipecolinic acid **190** from an aqueous solution of L-lysine **191**, as illustrated in Scheme 55. To achieve this photocatalytic transformation, the authors fabricated a Pyrex microreactor in which the channel cover plate was coated with a 300 nm layer of TiO$_2$ anatase (100 nm particles), prior to thermal bonding, to afford a titania-coated microreactor (TCM); the titania film was subsequently loaded with platinum (0.2 wt%), by photodeposition, to enable the TCM to be used for redox-combined photosynthesis. To conduct the reaction, the authors pumped a solution of L-lysine (**191**) (2.0 mM) through the reactor using a syringe pump and irradiated the TiO$_2$ film through a photomask using a high-pressure mercury lamp (110 mW cm^{-1}). The resulting reaction mixture was subsequently collected and analyzed offline by chiral HPLC in order to determine the proportion of L-lysine (**191**) converted to D- and L-pipecolinic acid **190/192**.

Employing a flow rate of 1 µl min^{-1} and a residence time of 0.86 min, the authors obtained 87% conversion of L-lysine (**191**), exhibiting 22% selectivity for pipecolinic acid and 14% yield of L-pipecolinic acid **190**. To demonstrate the TCM's efficiency, the authors also performed the reaction in batch employing 2 wt% Pt-loaded TiO$_2$ particles, which afforded the same surface to volume ratio of catalyst as calculated to be within the TCM, whereby a reaction time of 60 min afforded analogous results. The authors concluded that the increased reaction efficiency observed within the TCM was attributed to the efficient irradiation of the reaction mixture; however, for a true comparison they noted that measurement of the quantum yield of each system would be required.

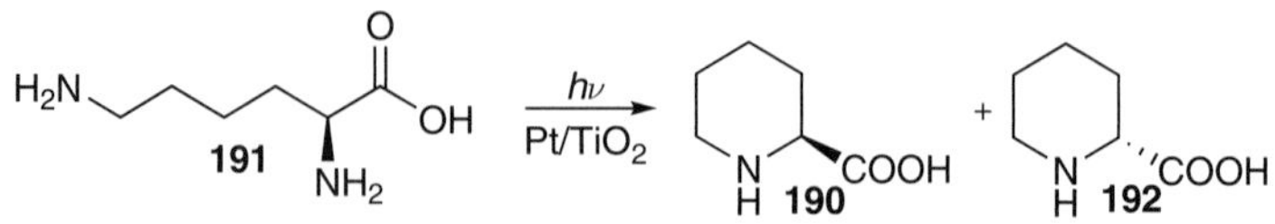

Scheme 55   Photocatalytic synthesis of L-pipecolinic acid **190** and D-pipecolinic acid **192**.

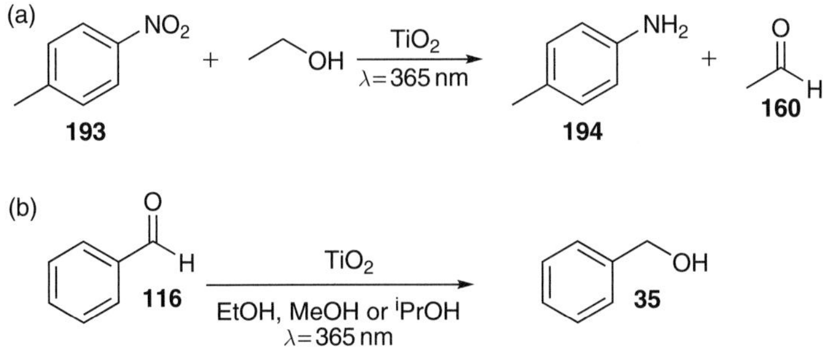

Scheme 56   Illustration of the photocatalytic reductions conducted by Matsushita et al. (2006) in a quartz microreactor.

In an analogous manner, Matsushita et al. (2006) investigated the use of a quartz microreactor [channel dimensions = 500 μm (wide) × 100 μm (deep) × 0.40 cm (long)], in which the bottom and sides of the microchannel were coated with TiO_2 (anatase), for the photocatalytic reduction of 4-nitrotoluene **193** (Scheme 56a) and benzaldehyde **116** (Scheme 56b) using a UV-LED ($\lambda = 365$ nm). As the photoreduction of an analyte requires a corresponding oxidation step to occur, a series of alcohols were evaluated as potential reaction solvents.

In theory, alcohols with low pK_a's afford a greater proportion of protons and alkoxy radicals, it was therefore proposed that MeOH would be an ideal solvent for this transformation; however, due to the kinetic instability of the methoxy radical the authors found the reductions to be more successful when conducted in EtOH. In addition, it was found that saturating the reactant stream with nitrogen, thus excluding dissolved oxygen, promoted the reaction as the electrons in the conduction band of the excited TiO_2 layer were not captured by oxygen. With this knowledge in hand, the authors evaluated the efficiency of the TiO_2-coated quartz reactor toward the reduction of 4-nitrotoluene **193** to afford 4-aminotoluene **194** and acetaldehyde **160**, as a function of reactant residence time within the microreactor. Employing 4-nitrotoluene **193** in N_2-saturated EtOH (1.0×10^{-4} M), in the absence of light, 0% conversion to 4-aminotoluene **194** was observed, this was subsequently increased to 8.3% with an irradiation time of 10 s and finally to 45.7% **194** when a residence time of 60 s was employed. Having demonstrated the ability to perform photocatalytic reduction of a nitro group to an amine, the authors investigated the reduction of benzaldehyde **116** (Scheme 56b). As the nitro group is more readily reduced than an aldehydic functionality, it came as no surprise to the authors that only 10.7% conversion to benzyl alcohol **35** was observed when employing an irradiation time of 60 s. Consequently, further studies are currently underway by the group in order to optimize the excitation wavelength and

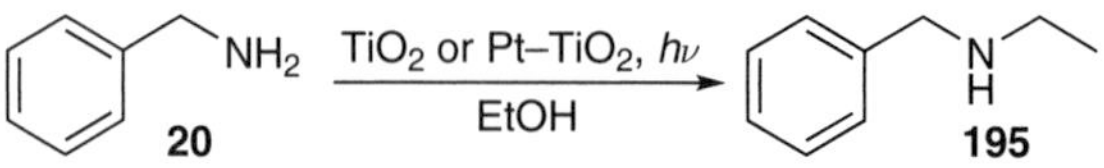

Scheme 57 The model reaction used to demonstrate the photochemical N-alkylation conducted in a quartz microreactor.

microreactor design for the reduction of multifunctional compounds (Matsushita et al., 2008).

Matsushita et al. (2007) subsequently demonstrated the ability to N-alkylate amines (Scheme 57) under continuous flow, again employing a quartz microreaction channel coated with a TiO_2 or Pt-loaded TiO_2 layer. As Table 28 illustrates, the illuminated specific surface area per unit of liquid attained within a microchannel is large even without taking into account the surface roughness of the catalyst; however, it can be seen that a shallow reaction channel provides optimal photon efficiency.

Having observed no photochemical alkylation in batch when employing TiO_2 particles and 84% N-alkylation (with 2.4% dialkylation after 5 h and 74.1% after 10 h) using the Pt–TiO_2 catalyst (5 h), initial flow reactions were conducted using a microreactor coated with Pt–TiO_2. Employing a series of UV-LEDs ($\lambda = 365$ nm, 2.2 m W cm^{-2}) as the light source, the authors pumped a solution of benzylamine **20** (1 mM) in EtOH through the Pt–TiO_2-coated microchannel and evaluated the effect of flow rate (2–$40\,\mu l\,min^{-1}$) and conversion to N-ethylbenzylamine **195**. Using a residence time of 2.5 min, the authors were pleased to observe 85% yield of N-ethylbenzylamine **195** with no sign of the dialkylated by-product. Having successfully performed a photochemical N-alkylation using the batch evaluated catalyst, the authors investigated the use of a Pt-free TiO_2-coated microchannel and were surprised to find that with an irradiation time of 90 s, 98% **195** was obtained. As Table 29 illustrates, increasing the surface to volume ratio by reducing the channel depth affords a more efficient reaction system and compared to batch where 0% **195** was obtained using TiO_2, excellent yields of N-ethylbenzylamine **195** can be

Table 29 Illustration of the effect of channel depth on the illuminated specific surface areas per unit of liquid in a microreactor (constant channel width of 500 μm) and the yield of N-ethylbenzylamine **195** obtained using a TiO_2 wall coating

Channel depth (μm)	Illuminated surface area ($m^2\,m^{-3}$)	Yield (%)
300	7.3×10^3	98
500	6.0×10^3	94
1,000	4.0×10^3	70

obtained with irradiation times as low as 1.5 min. It was also encouraging to see that the use of different solvent systems also enabled the authors to access *N*-methyl and *N*-propyl derivatives, providing a facile route to *N*-alkyl amines.

Wootton et al. (2002) investigated the use of a microreactor for the continuous photochemical generation of singlet oxygen and subsequently demonstrated the principle for the synthesis of ascaridole **196** (Scheme 58). To perform the reaction, a methanolic solution of α-terpinene **197** and Rose bengal (**198**) (photosensitizer) was introduced into the reactor at a flow rate of $1 \, \mu l \, min^{-1}$, where it was mixed with a constant stream of oxygen ($15 \, \mu l \, min^{-1}$), prior to irradiation with a 20 W tungsten lamp. As a safety precaution, the reaction products were degassed with nitrogen in order to prevent the accumulation of oxygenated solvents. Using the aforementioned protocol, the authors report 85% conversion of α-terpinene **197** to ascaridole **196**, demonstrating efficient photon transfer using a commercially available light source.

Another interesting photochemical transformation conducted in a microfabricated reactor, was the photochemical chlorination of toluene-2,4-diisocyanate reported by Ehrich et al. (2002). Employing a falling film microreactor [channel dimensions = 600 μm (wide) × 300 μm (deep) × 6.6 cm (long)], consisting of 32 parallel microchannels, the authors investigated the photochemical generation of chlorine radicals from gaseous chlorine using tetrachloroethane as the reaction solvent. By varying the flow rate of chlorine ($14–56 \, ml \, min^{-1}$) and toluene-2,4-diisocyanate ($0.1–0.6 \, ml \, min^{-1}$) feeds and the reactor temperature, the authors were able to optimize the production of benzyl chloride-2,4-diisocyanate. Employing a residence time of just 9 s and a reactor temperature of 130 °C, afforded benzyl chloride-2,4-diisocyanate in 81% yield with a space time yield of $401 \, mol \, h^{-1}$, which far exceeds the throughput of $1.3 \, mol \, h^{-1}$ attained in a conventional reactor.

See Section 3 for a discussion of the photochemical synthesis of an endothelin receptor antagonist **199** using the Barton reaction and the synthesis of a precursor to (−)-rose oxide **200** which is of industrial interest to fragrance manufacturers.

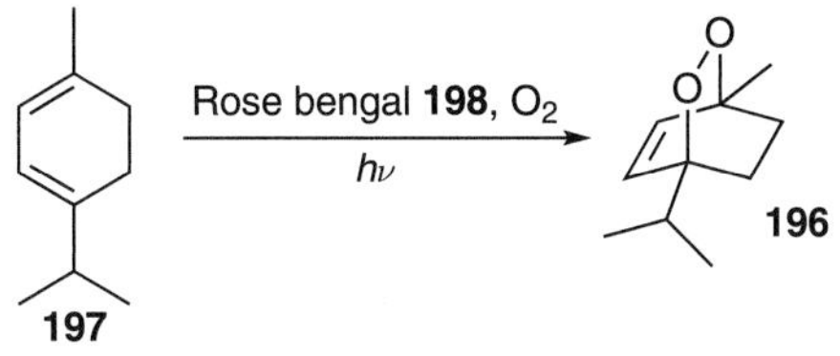

Scheme 58 An early example of the photochemical generation of singlet oxygen within a glass microreactor.

3. COMPOUNDS OF INTEREST AND INDUSTRIAL APPLICATIONS OF MICROREACTION TECHNOLOGY

Having discussed some of the advantages associated with the use of microreaction technology as a tool to conduct high-throughput organic synthesis, the final section of this chapter focuses on some of the compounds generated under continuous flow that are of synthetic and/or industrial interest.

Along with the halogen–lithium exchange reactions detailed in Section 2.3.3, numerous hydrogen–lithium exchange reactions have been reported to benefit from being conducted under continuous flow conditions. Such reactions are typically performed at low temperatures, that is, −78 °C, owing to the exothermic nature of the transformation and to prevent decomposition of unstable intermediates. The use of cryogenic reaction temperatures is however disadvantageous when considering employing reactions on a production scale. With this in mind, examples have featured within the literature that demonstrates the ability to increase reaction temperature, by conducting the reactions within miniaturized, continuous flow reactors. One such example of this is the synthesis of a sprio lactone **201** which is a fragment of neuropeptide Y, a receptor antagonist for the treatment of obesity (Takasuga et al., 2006).

As Scheme 59 illustrates, the reaction involves the hydrogen–lithium exchange of phenyl isonicotinamide **202**, in the presence of *n*-BuLi **74**/LiBr **203**, followed by a reaction with ethyl-4-oxocyclohexanecarboxylate **204** to afford the target spiro lactone **201** as a mixture of *cis/trans* isomers. In the presence of any residual dilithiated intermediate **205**, the spiro lactone **201** can undergo a second reaction to afford the by-product **206**. Conducting the reaction in a microflow reactor, comprising of static mixers and

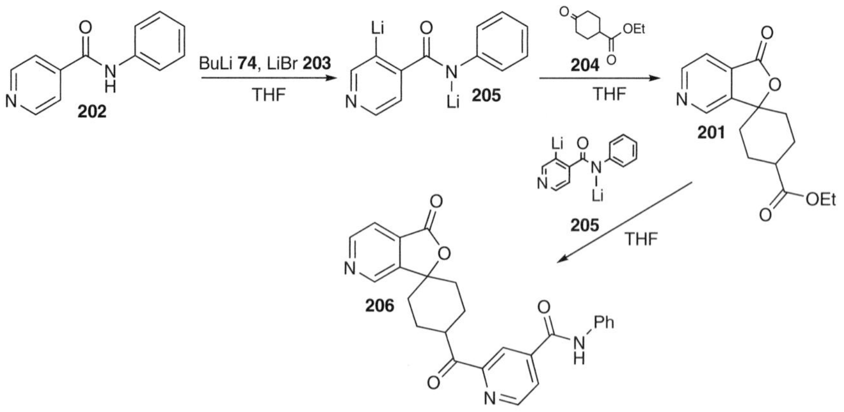

Scheme 59 Selective synthesis of a spiro lactone derivative **201** achieved using continuous flow methodology.

Table 30 Summary of the results obtained for the continuous flow synthesis of a spiro lactone **201**

Reactor temperature (°C)	Flow rate (ml min⁻¹)	Yield **201** (%)	Cis/ trans	Yield **206** (%)	Residual **202** (%)
−80	3.3	92.5	1.10	0.5	5.5
−80	3.3	93.3	1.12	0.8	6.9
−80	3.8	92.4	1.12	0.8	6.5
−10	5.3	51.8	1.75	32.5	10.1
−40	5.3	67.8	1.39	15.1	7.9
−80	7.2	90.9	1.18	1.9	8.1

stainless-steel tube reactors [0.25–2 mm (i.d.) × 0.01–20 m (length)], Takasuga et al. (2006) obtained yields in excess of 90% **201**; employing a reaction temperature of −80 °C and a reaction time estimated to be 20 s. The authors also noted that the yield decreased with increasing temperature and observed a change in selectivity upon varying the reactor temperature. In comparison, when increasing from an 800 tank to a 2,000 l tank (jacket temperature −90 °C, reaction mixture = −66 °C), the investigators observed a reduction in yield from 82.9 to 77.3% and an increase in the addition time from 15 to 36 min. The increased yield in the flow reactor can be attributed to the efficient removal of heat from the reaction mixture, preventing undesirable decomposition products obtained at increased temperatures, as illustrated in Table 30.

Compared with the above examples, whereby an array of pharmaceutically important molecules have been synthesized under pressure-driven flow, Garcia-Egido et al. (2002) reported the synthesis of fanetizole (**207**), an active compound for the treatment of rheumatoid arthritis, utilizing EOF. Employing a borosilicate glass microreactor fabricated at The University of Hull, the authors demonstrated the first example of a heated EOF-controlled reaction. As Scheme 60 illustrates, using

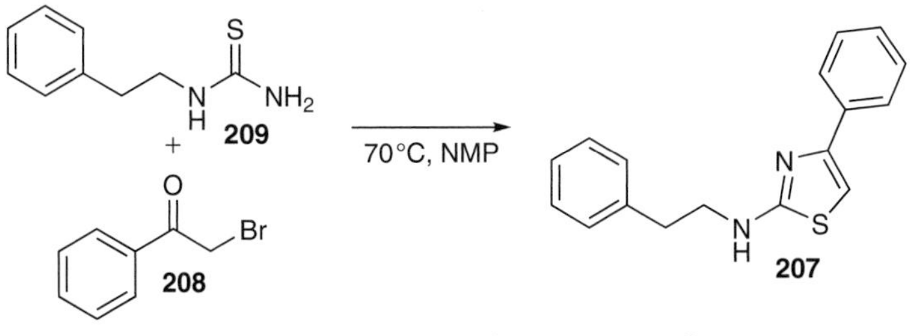

Scheme 60 Synthesis of fanetozole (**207**) under continuous flow.

N-methylpyrrolidone (NMP) as the reaction solvent, 2-bromoacetophenone **208** (1.4×10^{-2} M) and phenylethylthiourea **209** (2.1×10^{-2} M) were reacted to afford the target molecule **207** in quantitative conversion with respect to 2-bromoacetophenone **208**.

Employing a stainless-steel continuous flow reactor, Zhang et al. (2004) described the synthesis of gram to kilogram quantities of material for use in early clinical studies. One reaction reported by the authors was the exothermic synthesis of *N*-methoxycarbonyl-L-*tert*-leucine **210**, as illustrated in Scheme 61. By continuously adding a solution of methyl chloroformate **211** to L-*tert*-leucine **212**, in the presence of aq. NaOH **26**, at a reactor temperature of $-40\,^{\circ}$C afforded the target compound **210** in 91% yield with a throughput of $83.0\,\mathrm{g\,h^{-1}}$.

Using a stainless-steel microreactor, Ushiogi et al. (2007) reported a microflow system capable of synthesizing unsymmetrical diarylethenes, a process that is notoriously difficult to achieve in conventional batch systems. As Scheme 62 illustrates, diarylethenes are of synthetic interest due to their ability to change color via a reversible switching between two distinct isomers, which occurs as a result of light absorption.

With conventional protocols requiring low reaction temperatures, typically $-78\,^{\circ}$C, to prevent side reactions from occurring, scaling the reaction for industrial production of such compounds has proved difficult. As such, the authors evaluated the process under continuous flow, proposing that the effective temperature control and accurate residence times attainable within miniaturized flow reactors would enable the synthesis of diarylethenes at temperatures above $-78\,^{\circ}$C and thus facilitate the large-scale synthesis of such compounds.

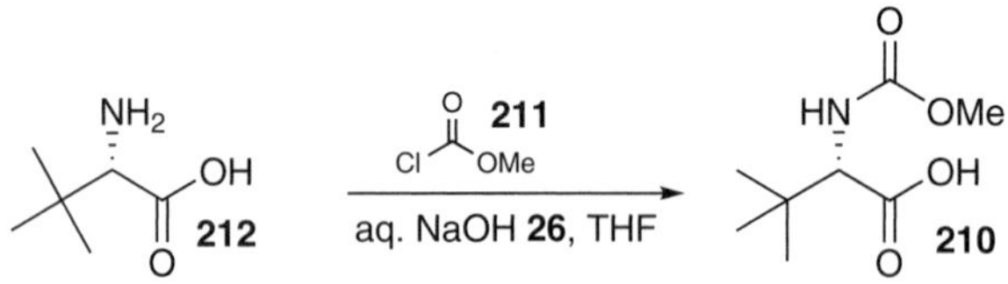

Scheme 61   An example of the reaction protocol employed for the exothermic synthesis of carbamates.

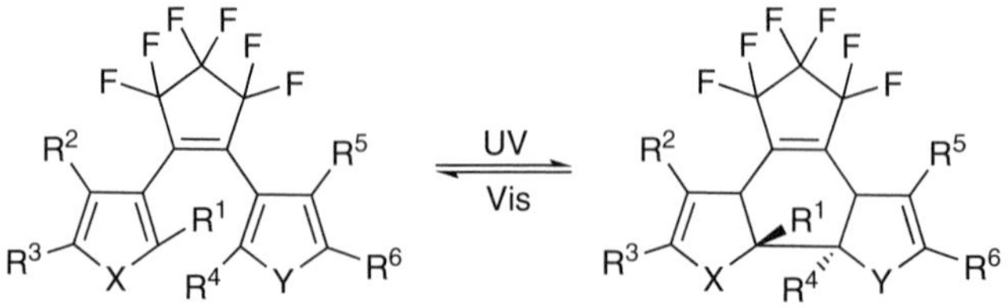

Scheme 62   Illustration of the photochromism exhibited by diarylethenes.

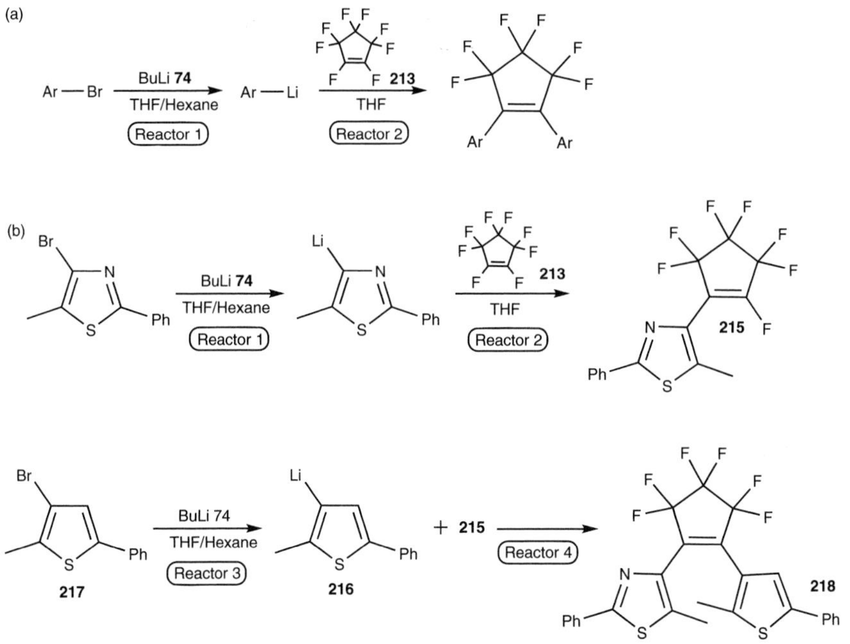

Scheme 63 Schematic detailing the general protocol employed for the continuous flow synthesis of (a) symmetrical and (b) unsymmetrical diarylethenes.

Initial investigations centered on the synthesis of symmetrical diary-lethenes and focused on identifying the effect of reactor temperature on the reaction yield. The first step of the reaction was a halogen–lithium exchange between an aryl bromide (0.3 M, 7.5 ml min^{-1}) and n-butyllithium **74** (1.5 M, 1.5 ml min^{-1}), traditionally performed at low reaction temperatures to avoid decomposition of the aryllithium intermediate (Scheme 63a). The second step involved the reaction of two equivalents of the aryllithium com-pound with octafluorocyclopentene **213** (0.75 M, 1.5 ml min^{-1}) to afford the target diarylethene **214**. Utilizing a stainless-steel microreactor con-sisting of two T-micromixers and two residence time units, the afore-mentioned reaction was conducted at a range of reaction temperatures (−45 to 15 °C) achieved by immersing the microreactor in a cooling bath, whereby 0 °C provided an optimum reaction temperature. Employing a residence time of 3.4 s in the first reactor and 2.9 s in the second, the authors isolated a range of diarylethenes in yields ranging from 47 to 87% depending on the aryl bromide used. In comparison to batch reac-tions conducted at −78 °C, this represented a dramatic increase in yield, with the monoarylated dominating in batch. The authors noted however that employing only 1 eq. of BuLi **74** within the flow reactor afforded the monoaryl derivative as the major product (Table 31). As such, it was

Table 31 Comparison of the product distribution obtained in batch and microflow systems for the arylation of octafluorocyclopentene **213**

		Isolated yield (%)	
Reactor	Reaction temperature (°C)	Diarylethene	Monoarylethene
Batch	−78 (3 h)[a]	10	52
Micro	0 (9 s)[a]	51	11

[a]The numbers in parentheses represent the reaction times employed.

proposed that by employing two different aryl halides, unsymmetrical diarylethenes could be obtained in high yield.

Using a convergent synthetic approach, the authors were able to adapt their flow process to enable the synthesis of unsymmetrical diarylethenes in good isolated yield. As depicted in Scheme 63b, the methodology involved preparing one monoarylated product **215** (reactors 1 and 2), as previously discussed, and the aryllithium intermediate **216** of a second aryl bromide **217** in parallel (reactor 3). The reactant streams were then converged in reactor 4, to afford the target unsymmetrical diarylethene **218** in an isolated yield of 53%. Owing to the different photochromic properties of such compounds, the synthetic methodology presented affords a facile route to the fine tuning of the compounds physical properties, whilst providing a method capable of producing any successful materials on a large scale.

Panke et al. (2003) also demonstrated enhanced reaction control, with respect to the temperature-sensitive synthesis of 2-methyl-4-nitro-5-propyl-2H-pyrazole-3-carboxylic acid **219**, a key intermediate in the synthesis of the lifestyle drug Sildenafil® (**220**) (Scheme 64). When performing the nitration of 2-methyl-5-propyl-2H-pyrazole-3-carboxylic acid **219** under adiabatic conditions, with a dilution of 6.0 l kg^{-1}), Dale et al. (2000) observed a temperature rise of 42 °C (from 50 to 92 °C) upon addition of the nitrating solution. As Scheme 63 illustrates, this proved problematic as at 100 °C decomposition of the product **219** was observed and in order to reduce thermal decomposition of pyrazole **219**, and increase process safety, the authors investigated addition of the nitrating solution in three aliquots, which resulted in a reduced reaction temperature of 71 °C and an increase in chemoselectivity; unfortunately, the reaction time was increased from 8 to 10 h.

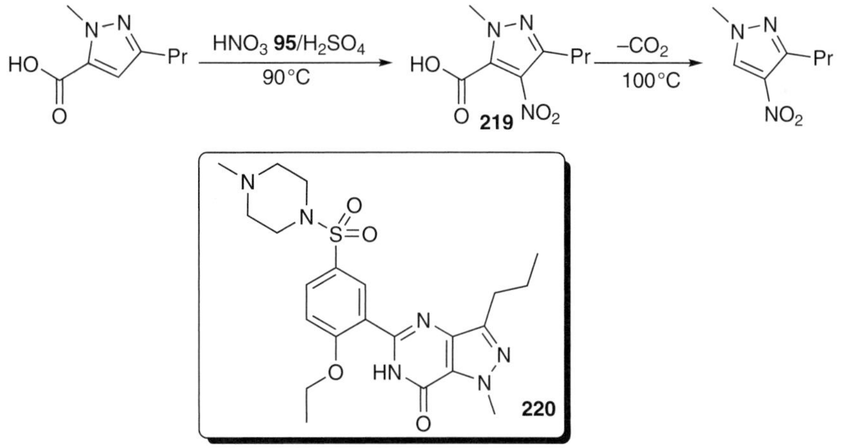

Scheme 64 An illustration of the temperature-sensitive synthesis of 2-methyl-4-nitro-5-propyl-2H-pyrazole-3-carboxylic acid **219**, a key intermediate of Sildenafil® (**220**).

By conducting the reaction in a flow reactor, where the heat of reaction can be rapidly dissipated, the authors were able to maintain a reaction temperature of 90 °C as a result of adding the nitrating mixture continuously. Coupled with a residence time of 35 min, the authors were able to attain a throughput of $5.5 \, g \, h^{-1}$ with an overall yield of 73% **219**. In addition to the dramatic reduction in residence time (10 h–35 min) and the increased process safety, the continuous flow methodology afforded a facile route to the chemoselective synthesis of 2-methyl-4-nitro-5-propyl-2H-pyrazole-3-carboxylic acid **219**.

Having earlier demonstrated the use of AlMe$_3$ **71** for the efficient synthesis of amides from a series of simple methyl and ethyl esters, Gustafsson et al. (2008b) extended their investigation to include the synthesis of two pharmaceutically relevant compounds, rimonabant (**72**) (SR141716) (Scheme 65) and efaproxiral (RSR13) (**73**) (Scheme 66). In both cases, the established synthetic protocols involved the formation of an amide bond, in the first instance from and acid chloride and 1-aminopiperidine **221** and secondly from the reaction of the ester or carboxylic acid and a substituted aniline precursor. Based on their previous success employing AlMe$_3$ **71** in a PTFE tubing (1 mm i.d.)-based continuous flow reactor, the authors investigated the synthesis of rimonabant (**72**) and efaproxiral (**73**) under continuous flow conditions.

As Scheme 65 illustrates, the first step of the rimonabant (**72**) flow synthesis involved the treatment of 4-chloropropiophenone **222** with LiHMDS **223** for 1 min prior to the addition of ethyl oxalate **224** at 50 °C where it reacted for 5 min; after work up and purification, the resulting β-ketoester **225** was isolated in 70% yield. Treatment of **225** with the HCl

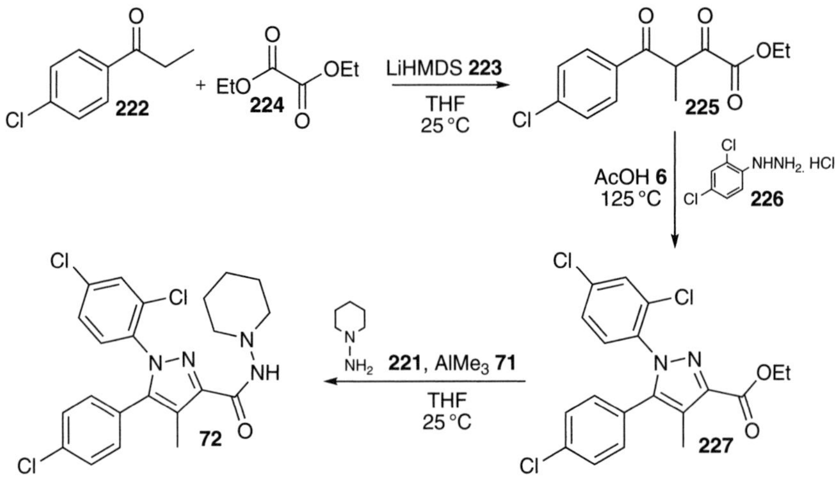

Scheme 65 Synthetic route employed for the continuous flow synthesis of rimonabant (72).

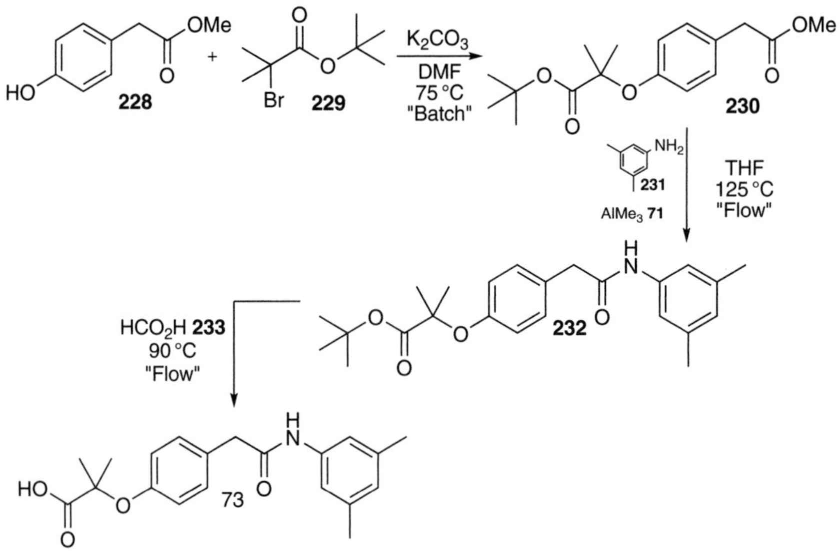

Scheme 66 Combination of batch and flow methodology for the synthesis of the pharmaceutical agent, efaproxiral (73).

salt of 4-chlorophenylhydrazine 226, in acetic acid 6, at 125 °C afforded pyrazole 227 in 80% yield with a residence time of 16 min. The final step of the synthesis involved the formation of an amide bond which was achieved via the treatment of pyrazole 227 with AlMe₃ 71 and

1-aminopiperidine **221** in THF at 125 °C for 2 min. Using this approach, the authors reported the isolation of the antiobesity drug rimonabant (**72**) in an overall yield of 49%, demonstrating the ability to continuously synthesize drug molecules on the gram scale.

In synthesis of their second drug target, Seeberger et al. combined batch and flow regimes as a means of obtaining efaproxiral (**73**), a pharmaceutical agent used for the enhancement of radiation therapy. This mode of operation was selected as the authors acknowledged potential problems associated with the heterogeneous nature of the inorganic base/ organic solvent mixture that was required to alkylate the phenol derivative **228** (Scheme 66). In batch, the authors alkylated phenol **228** with the *tert*-butyl ester of 2-bromo-2-methyl-propionic acid **229** to afford the methyl ester **230** in 75% yield. The second step of the reaction involved an aluminum-mediated amide bond formation between the ester **230** and 3,5-dimethylaniline **231** which was conducted under continuous flow, using a residence time of 2 min, affording the amide **232** in 77% yield and quantitative selectivity toward the methyl ester. The final step involved hydrolysis of the *tert*-butyl ester, which was achieved under flow using formic acid **233** at 90 °C, affording efaproxiral (**73**) in 89% yield. Using this combination of batch and flow reactions, the authors were able to synthesize the target compound **73** at 24 mmol h^{-1}.

In a recent example of industrial interest, Tietze and Liu (2008) reported the development of a continuous flow process for the synthesis of an aminonaphthalene derivative **91** on a kilogram scale, for use as a starting material toward the preparation of novel anticancer agents such as the duocarmycin prodrug **234** (Scheme 67). Initial investigations focused on modifying the existing batch procedure and utilizing numerous single microreaction steps in series, thus enabling comparison of the developed flow protocol with existing synthetic methodology.

As Scheme 68 illustrates, the first step of the reaction was the synthesis of the *tert*-butyl ester **235** from bromoacetyl bromide **236**. Owing to the

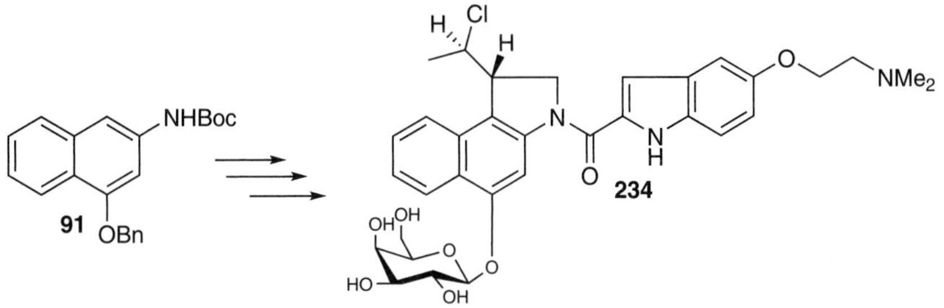

Scheme 67   Schematic illustrating compound **91** as a building block in the synthesis of the prodrug **234**.

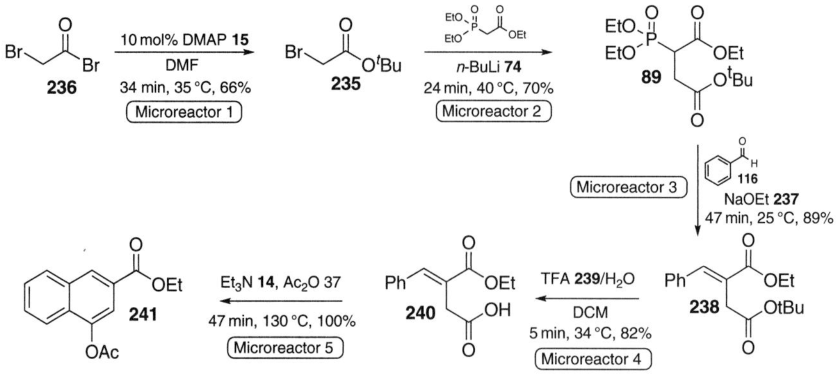

Scheme 68 Schematic illustrating the first five microreactions used in the synthesis of **91**.

precipitation of an ammonium salt, solvents such as diethyl ether and THF were unsuitable for use within the flow reactor; a short solvent study was therefore conducted and DMF identified as the ideal solvent. Employing 10% DMAP **15** in DMF, a reaction temperature of 35 °C and a residence time of 34 min, the authors were able to isolate the target *tert*-butyl ester **235** in 66% yield. The *tert*-butyl ester **235** was subsequently used in the synthesis of the phosphonosuccinate **89**; details discussed previously and can be found summarized in Scheme 23.

In a third microreactor, the anion of 4-*tert*-butyl 1-ethyl-2-(diethox-yphosphoryl)succinate was prepared *in situ* using sodium ethoxide **237** (in EtOH) and the Wittig–Horner olefination with benzaldehyde **116** performed using a residence time of 47 min to afford (*E*)-*tert*-butyl-1-ethyl-2-benzylidenesuccinate **238** in excellent selectivity (89% yield). In a fourth reactor, the acid-catalyzed (TFA **239**) *tert*-butyl ester depro-tection was achieved using a residence time of 5 min at 34 °C and employing DCM as the reaction solvent to afford (*E*)-3-(ethoxycarbonyl)-4-phenylbut-3-enoic acid **246** in 82% yield. The deprotection was subse-quently followed by a Friedel–Crafts acylation, using triethylamine **14** and acetic anhydride **37**, to afford 4-acetoxy-naphthalene-2-carboxylic acid ethyl ester **241** in quantitative yield when conducted at 130 °C (residence time = 47 min).

In the first step illustrated in Scheme 69, the carboxylic acid ethyl ester **241** undergoes quantitative solvolysis, using NaOEt **237** (30 mol%), the resulting naphthol derivative **242** was subsequently protected as its benzyl ether **244** using aq. NaOH **26** and benzyl bromide **46** to afford ethyl-4-(benzyloxy)-2-naphthoate **243** in 72% yield. Quantitative hydrolysis of the ethyl ester **243** followed, using aq. NaOH **26** at 68 °C for 48 min. The carboxylic acid **244** was used directly with the Shioiri–Yamada reagent

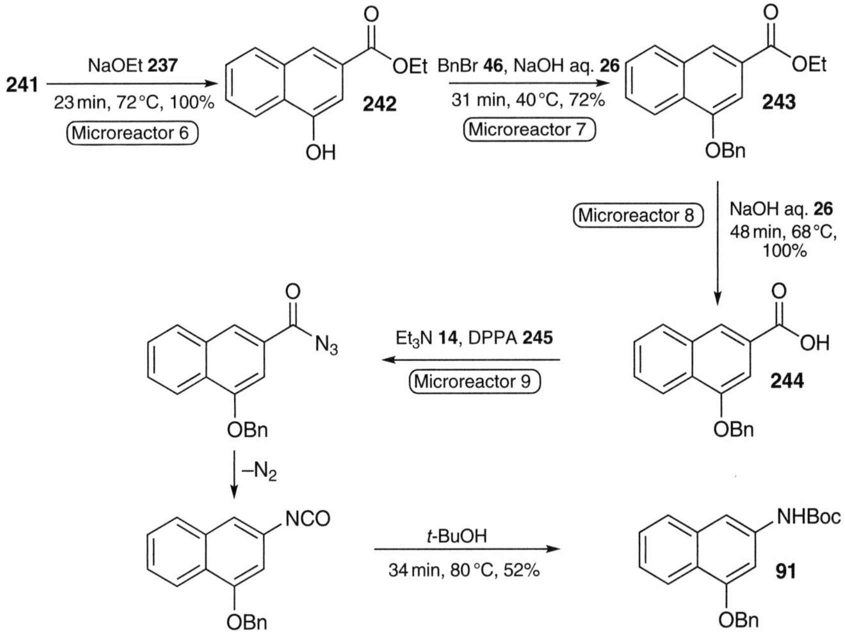

Scheme 69 Schematic illustrating the final four microreactions used to obtain the target (4-benzyloxynaphthalen-2-yl)-carbamic acid *tert*-butyl ester **91**.

DPPA **245** in the presence of *tert*-butanol to afford the target product **91** via a Curtius rearrangement (52% yield).

The authors were pleased to report that in most cases, similar or better results were obtained when comparing the reaction steps performed under continuous flow and batch, the continuous flow approach, however, had the advantages of providing increased safety and faster reactions, with an empirical accelerating factor of $F = 3$–10 reported.

In an analogous manner, LaPorte et al. (2008) employed three consecutive reactions in the development of a continuous protocol for the synthesis of 6-hydroxybuspirone **246** (Scheme 70). The authors rationalized the need for a continuous process as the oxidation times required at a pilot plant stage would be of the order of 16–24 h as the process was mass transfer limited and the need to cool the reaction mixture to $-70\,^\circ$C would be an issue for concern. To simplify studies, the authors employed a solution of the preformed enolate **247** (prepared at $-70\,^\circ$C using NaHMDS **44**, 3 ml min^{-1}) as the reactant feedstock for one inlet of the microreactor and oxygen (0.3 l min^{-1}) as the second feed. The microreactor was cooled to $-10\,^\circ$C using a recirculating coolant and afforded 65–70% conversion to **246** with a residence time of 2–3 min. To increase the yield of the target

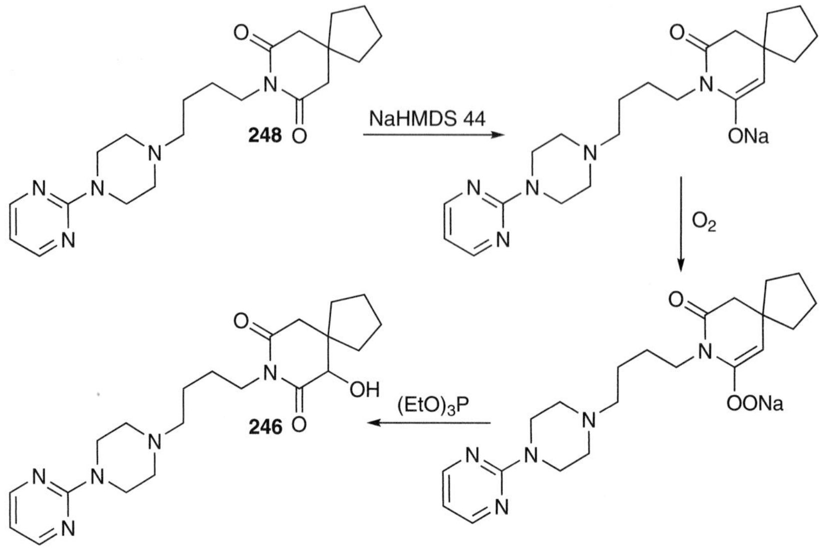

Scheme 70 Synthesis of 6-hydroxybuspirone **246** via the hydroxylation of the azapirone psychotropic agent buspirone (**248**).

product further, the authors subsequently employed a second microreactor where the reaction products from the initial reactor were reacted with a second stream of oxygen. Using this approach, the yield was increased to 85–92% with a processing time of 5–6 min with a daily production volume of 300 g day^{-1}. Based on the encouraging results obtained, the process was subsequently scaled to enable the production of 6-hydroxybuspirone **246** from buspirone (**248**) on a multi-kilo scale and incorporated in-line FTIR monitoring.

In an extension to their earlier examples of β-hydroxyketone dehydration (Scheme 7), Tanaka et al. (2007) evaluated the continuous flow synthesis of an immunoactivating natural product, pristane (2,6,10,14-tetramethylpentadecane) (**249**). Due to limited commercial availability of pristane (**249**), the authors investigated the compounds preparation in a microreactor as a means of obtaining a method suitable for production of pristane (**249**) to meet demand, which is currently ~5 kg week^{-1}. The first step of the reaction involved treating farnesol (**250**) with MnO$_2$ to afford the respective aldehyde, which subsequently underwent a reaction with isobutyl magnesium chloride to afford an allylic alcohol **251**. The alcohol **251** (1.0 M in THF) was subsequently dehydrated within a micromixer (Comet X-01) using p-TsOH·H$_2$O **12** in THF/toluene (0.2–1.0 M), at total flow rate of 600 μl min^{-1} and a reaction temperature of 90 °C, followed by catalytic hydrogenation. Under the aforementioned reaction conditions, the authors obtained the target compound **249** in 80% yield from farnesol

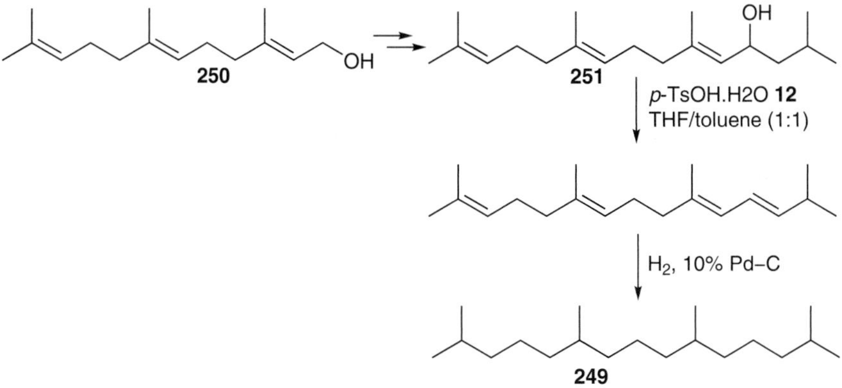

Scheme 71 Synthetic strategy employed for the continuous flow synthesis of the immunoactivating natural product, pristane (**249**).

(**250**) (Scheme 71). Compared to batch techniques, this synthetic route proved advantageous as only a simple purification was required in order to isolate the product **249** unlike the multiple distillations traditionally employed.

Cheng-Lee et al. (2005) demonstrated the multistep synthesis of a radiolabeled imaging probe in a PDMS microreactor, consisting of a complex array of reaction channels, with typical dimensions of 200 μm (wide) 45 μm (deep). Employing a sequence of five steps, comprising of (1) [^{18}F] fluoride concentration (500 μCi), (2) solvent exchange from H_2O to MeCN, (3) [^{18}F]fluoride substitution of the D-mannose triflate **252** (324 ng), to afford the labeled probe **253** (100 °C for 30 s and 120 °C for 50 s), (4) solvent exchange from MeCN to H_2O, and finally, (5) acid hydrolysis of **254** at 60 °C, the authors demonstrated the synthesis of 2-[^{18}F]-FDG **254** (Scheme 72).

Using this approach, 2-[^{18}F]-FDG **254** was obtained in 38% radiochemical yield, with a purity of 97.6% (determined by radio TLC). Miniaturization of the process enabled the reaction sequence to be performed in

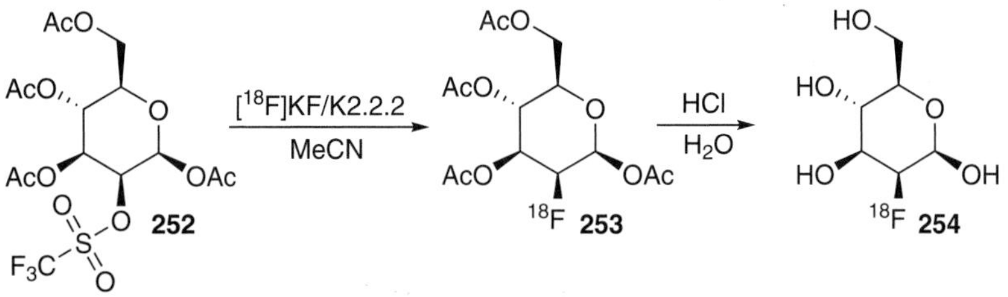

Scheme 72 Synthesis of the radiolabel 2-[^{18}F]-fluorodeoxyglucose (2-[^{18}F]-FDG) **254**.

14 min, compared to 50 min for the current automated protocol. Most importantly, however, the authors found the process to be reproducible between runs and reactors. Future work is therefore concerned with increasing the reaction chamber in order to increase the quantities of material synthesized for use in human PET imaging (10 mCi patient^{-1}).

Building on the bifurcated pathway, developed to enable the selective synthesis of thiazoles or imidazoles (Scheme 73), Baxendale et al. (2005) subsequently demonstrated the synthesis of a HIV-1 RTI analog **255** using the same reaction methodology.

As Scheme 73 illustrates to attain the target compound **255**, the authors reacted 4-bromophenyl isothiocyanate **256** with ethyl isocyanoacetate **257** in the presence of PS-BEMP **124** to afford the thiazole **258** in 58% yield and introduction of α-bromoamide **259** through the PS-BEMP **124** cartridge subsequently afforded the target compound **255** in 30% yield.

In 2005, Baxendale et al. reported the first enantioselective synthesis of 2-aryl-2,3-dihydro-3-benzofurancarboxamide neolignan (grossamide) **260** conducted under continuous flow conditions. As illustrated in Scheme 73, the first step of the reaction involved the synthesis of amide **261** *via* the coupling of ferulic acid **262** and tyramine **171**, in the presence of PS-HOBt **263**. Monitoring reaction progress by LC–MS, the authors were able to optimize this step to afford the amide **261** in 90% conversion; however, prior to performing the second reaction step it was imperative to remove any residual tyramine **171**. As Scheme 74 illustrates, this was achieved by

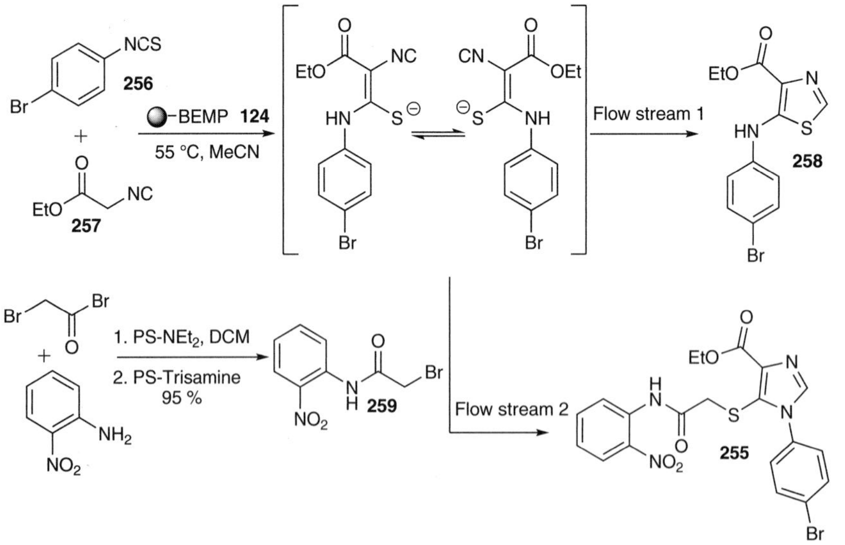

Scheme 73   The continuous flow synthesis of a HIV-1 RTI analog **255** from α-bromoamide **259**.

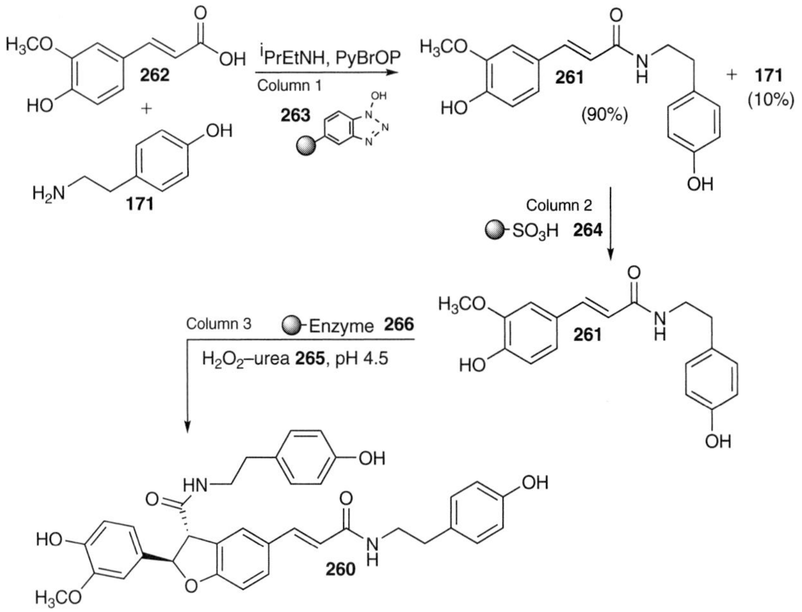

Scheme 74 Continuous flow protocol employed for the synthesis of the natural product grossamide (**260**) using a series of solid-supported reagents, catalysts and scavengers.

passing the reaction mixture through a second column containing PS-SO$_3$H **264** to afford the target intermediate **261** in excellent purity. A premixed solution of the amide **261** and H$_2$O$_2$–urea complex **265**, in aqueous buffer, was subsequently pumped through a third column, containing silica-supported peroxidase **266**, to afford grossamide (**260**) in excellent yield and purity.

Employing a series of packed columns, containing immobilized reagents, catalysts, and scavengers, in conjunction with a glass microreactor and a flow hydrogenation system, Baxendale et al. (2006) reported the flow-assisted synthesis of the alkaloid natural product ($\pm$)-oxomaritidine (**266**). As depicted in Scheme 75, the initial steps of azide **267** and aldehyde **268** synthesis were performed simultaneously in separate reaction columns and at the point of convergence, in the presence of a polymer-supported phosphine **269** at 55 °C, afforded the aza-Wittig intermediate **270**. This was followed by reduction of the imine **270**, utilizing a commercially available flow hydrogenator containing 10% Pd/C, to afford the 2° amine **271** in THF. As subsequent reaction steps required a change of reaction solvent, from THF to DCM, the amine **271** was collected on-line and solvent removal achieved using a Vapourtec V-10 solvent evaporator, the amine **271** was then redissolved in DCM, a process that took less than

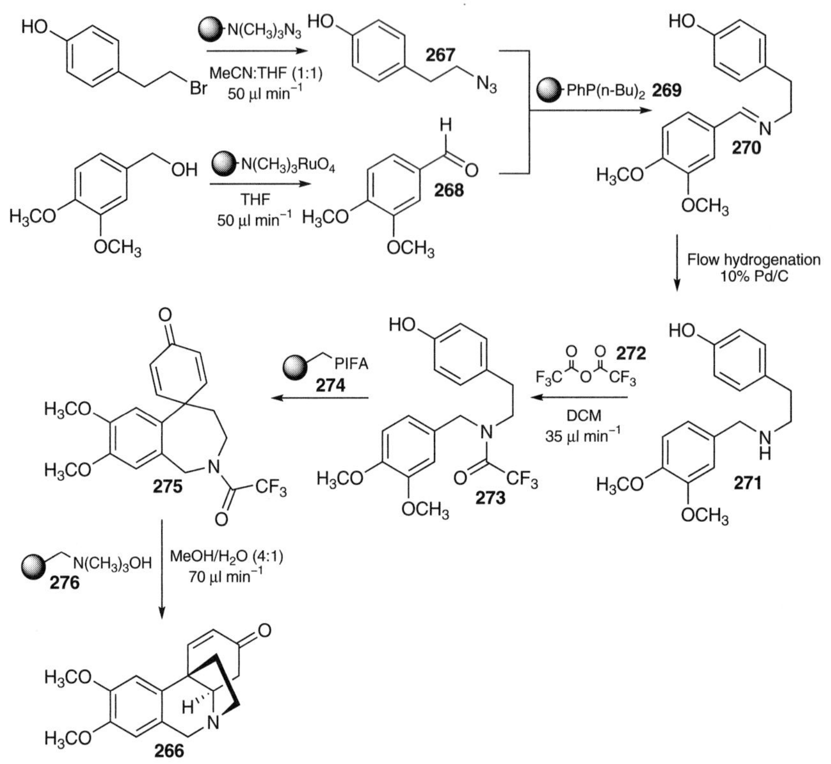

Scheme 75 Illustration of the reaction steps employed in the flow-assisted synthesis of (±)-oxomaritidine (**266**).

10 min. In the latter half of the reaction sequence, the 2° amine **271** was trifluoroacetylated (5 eq. of TFAA **272** 80 °C) within a glass microreactor to afford amide **273**, which subsequently underwent oxidative phenolic coupling, in the presence of polymer-supported (ditrifluoroacetoxyiodo)benzene **274**, to afford a seven-membered tricyclic derivative **275**. The resulting product stream was finally passed through a column reactor, containing a polymer-supported base **276**, which promoted cleavage of the amide bond followed by spontaneous 1,4-conjugate addition to generate the target compound (±)-oxomaritidine (**266**).

Employing a stainless-steel microreactor, Sugimoto et al. (2006, 2008) demonstrated the Barton reaction (nitrite photolysis) of a steroidal substrate **277** to afford **199**, a key intermediate in the synthesis of an endothelin receptor antagonist. Initial investigations employed a conventional 300 W high-pressure mercury lamp and a stainless-steel microreactor with a single serpentine reaction channel [1,000 μm (wide) × 107 μm (deep) 2.2 m (long)]. To perform a continuous Barton photolysis, a gap of 7.5 cm was maintained between the light source and the stainless-steel micro-reactor, an acetone solution containing the nitrite **277** (9 mM) and pyridine (0.2 eq.) was then pumped through the stainless-steel reactor at a flow rate of 33 µl min^{-1}, and the reaction products analyzed off-line by HPLC to determine the conversion of nitrite **277** to oxime **199**.

Using a quartz cover plate, the authors observed the production of a complex reaction mixture due to the low wavelength of the mercury light source. In comparison, using a Pyrex cover plate afforded the rearranged product **199** in moderate yield (21%), whereas employing a soda-lime glass cover plate increased the yield to 59% **199**. The authors found that if the light source was positioned closer to the reactor that is 5.0 cm, the yield of **199** decreased (33%), an observation that was attributed to thermal degradation of the oxime **199** occurring as a result of the light source heating the reactor (>50 °C) (Scheme 76).

As the short wavelength light (<365 nm) emitted from the mercury lamp was not thought to be advantageous for this particular reaction, the authors replaced the light source with a 15 W black light ($\lambda_{max} = 352$ nm) as a means of avoiding the undesirable heating of the microreactor. Due to the reduced heating of the light source, the black light was able to be positioned 3.0 cm from the reactor (cf. 7.5 cm for the mercury lamp). As Table 32 illustrates, a Pyrex cover plate afforded the highest yield of oxime **199** (21%) with a marked reduction in energy consumption (cf. to the mercury lamp) and the authors found that by increasing the reactant

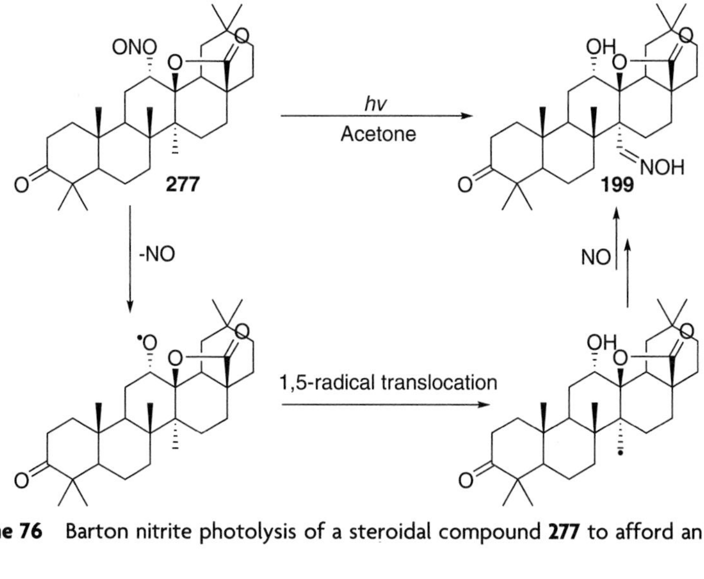

Scheme 76 Barton nitrite photolysis of a steroidal compound **277** to afford an oxime **199**.

Table 32 Summary of the results obtained during the Barton photolysis of **277**

Light source	Cover plate material	Residence time (min)	Yield (%)	Yield (W h^{-1})
300 W Hg lamp	Pyrex	6	21	0.7
300 W Hg lamp	Soda lime	6	56	1.9
15 W black light	Soda lime	6	15	10.3
15 W black light	Pyrex	6	29	19.3
15 W black light	Pyrex	12	71	23.7

residence time from 6 to 12 min, the target oxime **199** could be obtained in 71% yield; as determined by HPLC analysis.

In order to produce oxime **199** in greater quantities, the authors subsequently evaluated the use of DMF as the reaction solvent due to the increased solubility of the nitrite precursor **277** (36 mM). In conjunction with two serially connected microreactors, each containing 16 microchannels [1,000 μm (wide) × 500 μm (deep) × 1.0 m (length)] and eight black lights, the photochemical synthesis was performed continuously for 20 h at a flow rate of 250 μl min^{-1} (residence time = 32 min). After an off-line aqueous extraction and silica gel column, 3.1 g of the oxime **199** was obtained equating to an isolated yield of 60% and successfully demonstrating the ability to use photochemical synthesis for the scalable preparation of pharmaceutically relevant compounds.

A more recent example of the *in situ* photochemical generation of singlet oxygen was reported by Meyer et al. (2007) who demonstrated the synthesis of hydroperoxide **278**, a precursor to (–)-rose oxide (**200**) which is used as a fragrance in the perfume industry (Scheme 77). To demonstrate the synthetic utility of the continuous flow methodology, batch reactions were performed alongside the microreactions, using a modified Schlenk reactor (40 ml). The microreactor employed was fabricated from Borofloat$^®$ glass, possessed a total internal volume of 270 μl, and was illuminated using a diode array consisting of 4 × 10 diodes (λ 468 nm). To evaluate the microreactor, a solution of (–)-β-citronellol (**279**) (0.1 M) and Ru(^{t}bpy)$_3$Cl$_2$ **280** (1 × 10^{-3} M) in EtOH was firstly purged with compressed air for 20 min (0.4 l h^{-1}), using a peristaltic pump, the reaction mixture was subsequently cycled through the reactor over the course of the investigation, with samples analyzed off-line periodically by HPLC. In comparison to the batch reactor where only 7.5 ml of the reactor volume was illuminated (15.2 cm^2), in the microreactor all 270 μl of the reaction mixture was irradiated (6.9 cm^2) which afforded increased photonic efficiency leading to enhanced space time yields of 0.9 mmol l^{-1} min^{-1} (cf. 0.1 mmol l^{-1} min^{-1}) obtained in the batch cell.

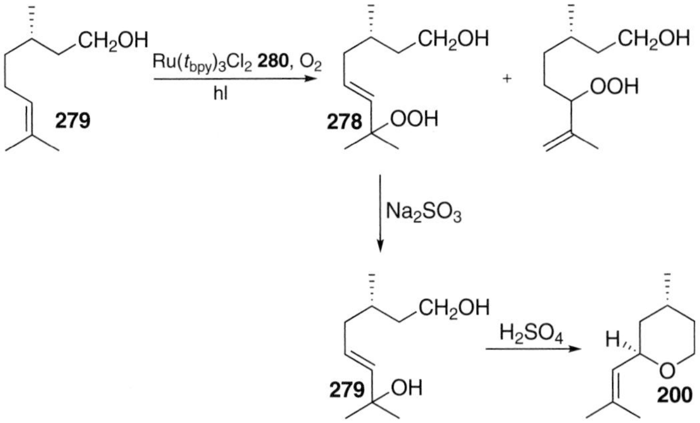

Scheme 77 Photooxygenation of (–)-β-citronellol (**279**) conducted in a glass microreactor and the subsequent use of hydroperoxide **278** in the synthesis of (–)-rose oxide (**200**).

4. CONCLUSIONS

As can be seen from the variety of reactions discussed herein, microreaction technology enables the rapid screening of both reactants and reaction conditions, thus affording a facile route to the generation of compound libraries. In addition to the time savings harnessed, the use of MRT has the potential to reduce the costs associated with high-throughput chemistry as it enables a vast array of information to be generated from reduced quantities of substrates and catalysts when compared with conventional techniques. Added to this the advantage that the reaction conditions identified can subsequently be transferred from the research laboratory to pilot-scale production and beyond, it is clear to see why the field of microreactor research has grown rapidly over the past decade.

REFERENCES

Acke, D. R., and Stevens, C. *Green Chem.* **9**, 386–390 (2007).

Acke, D. R., Stevens, C. V., and Roman, B. I. *Org. Proc. Res. Dev.* **12**, 921–928 (2008).

Annis, D. A., and Jacobsen, E. N. *J. Am. Chem. Soc.* **121**, 4147–4154 (1999).

Baumann, M., Baxendale, I. R., Ley, S. V., Nikbin, N., and Smith, C. D. *Org. Biomol. Chem.* **6**, 1587–1593 (2008).

Baumann, M., Baxendale, I. R., Ley, S. V., Nikbin, N., Smith, C. D., and Tierney, J. P. *Org. Bio. Mol. Chem.* **6**, 1577–1586 (2008b).

Baxendale, I. R., Deeley, J., Griffiths-Jones, C. M., Ley, S. V., Saaby, S., and Tranmer, G. K. *Chem. Commun.* 2566–2568 (2006).

Baxendale, I. R., Griffiths-Jones, C. M., Ley, S. V., and Tranmer, G. K. *Synlett* **3**, 427–430 (2005).

Baxendale, I. R., Ley, S. V., Smith, C. D., Tamborini, L., and Voica, A. *J. Combin. Chem.* **10**(6), 851–857 (2008).

Belder, D., Ludwig, M., Wang, L., and Reetz, M. T. *Angew. Chem. Int. Ed.* **45**, 2463–2466 (2006).

Benali, O., Deal, M., Farrant, E., Tapolczay, D., and Wheeler, R. *Org. Proc. Res. Dev.* **12**, 1007–1011 (2008).

Bogdan, A. R., Mason, B. P., Sylvester, K. T., and McQuade, D. T. *Angew. Chem. Int. Ed.* **46**, 1698–1701 (2007).

Braune, S., Pochlauer, P., Reintjens, R., Steinhofer, S., Winter, M., Lobet, O., Guidat, R., Woehl, P., and Guermeur, C. *Chem. Today* **26**, 1–4 (2008).

Burns, J. R., and Ramshaw, C. *Chem. Eng. Commun.* **189**, 1611–1628 (2002).

Carrel, F. R., Geyer, K., Codee, J. D.C., and Seeberger, P. H. *Org. Lett.* **9**(12), 2285–2288 (2007).

Chambers, R. D., Fox, M. A., Holling, D., Nakano, T., Okazoe, T., and Sandford, G. *Lab Chip* **5**, 191–198 (2005).

Chambers, R. D., Holling, D., Spink, R. C.H., and Sandford, G. *Lab Chip* **1**, 132–137 (2001).

Chambers, R. D., Sandford, G., Trmcic, J., and Okazoe, T. *Org. Proc. Res. Dev.* **12**(2), 339–344 (2008).

Chambers, R. D., and Spink, R. C.H. *Chem. Commun.* 883–884 (1999).

Chatgilialoglu, C. *Chem. Eur. J.* **14**, 2310–2320 (2008).

Cheng-Lee, C., Sui, G. D., Elizarov, A., Shu, C. Y.J., Shin, Y. S., Dooley, A. N., Huang, J., Daridon, A., Wyatt, P., Stout, D., Kolb, H. C., Witte, O. N., Satyamurthy, N., Heath, J. R., Phelps, M. E., Quake, S. R., and Tseng, H. R. *Science* **310**, 1793–1796 (2005).

Comer, E., and Organ, M. G. *Chem. Eur. J.* **44**, 7223–7227 (2005a).

Comer, E., and Organ, M. G. *J. Am. Chem. Soc.* **127**, 8160–8167 (2005b).

Costantini, F., Bula, W. P., Salvio, R., Huskens, J., Gardeniers, H. J.G.E., Reinhoudt, D. N., and Verboom, W. *J. Am. Chem. Soc.* **131**, 1650–1651 (2009).

Csajagi, C., Szatzker, G., Toke, R. R., Urge, L., Davas, F., and Poppe, L. *Tetrahedron: Asymmetry* **19**, 237–246 (2008).

Dale, J. D., Dunn, P. J., Golightly, C., Hughes, M. L., Levett, P. C., Pearce, A. K., Searle, P. M., and Ward, G. *Org. Proc. Res. Dev.* **4**, 17–22 (2000).

Drager, G., Kiss, C., Kunz, Y., and Kirschning, A. *Org. Biomol. Chem.* **5**, 3659–3664 (2007).

Ducry, L., and Roberge, D. M. *Angew. Chem. Int. Ed.* **44**, 7972–7975 (2005).

Ehrich, H., Linke, D., Morgenschweis, K., Baerns, M., and Jahnisch, K. *Chimia* **56**, 647–653 (2002).

Fernandez-Suarez, M., Wong, S. Y.F., and Warrington, B. H. *Lab Chip* **2**, 170–174 (2002).

Flogel, O., Codee, J. D.C., Seebach, D., and Seeberger, P. H. *Angew. Chem.* **45**, 7000–7003 (2006).

Fukuyama, T., Hino, Y., Kamata, N., and Ryu, I. *Chem. Lett.* **34**, 66–67 (2004).

Garcia-Egido, E., Wong, S. Y.F., and Warrington, B. H. *Lab Chip* **2**, 31–33 (2002).

Geyer, K., and Seeberger, P. H. *Helv. Chim. Acta.* **90**, 395–403 (2007).

Glasnov, T. N., and Kappe, C. O. *Macromol. Rapid Commun.* **28**, 395–410 (2007).

Gorges, R., Meyer, S., and Kreisel, G. *J. Photochem., Photobiol. A: Chem.* **167**, 95–99 (2004).

Goto, S., Velder, J., El Sheikh, S., Sakamoto, Y., Mitani, M., Elmas, S., Adler, A., Becker, A., Neudorfl, J.-M., Lex, J., and Schmalz, H.-G. *Synletters* **9**, 1361–1365 (2008).

Gustafsson, T., Gilmor, R., and Seeberger, P. H. *Chem. Commun.* 3022–3024 (2008a).

Gustafsson, T., Ponten, F., and Seeberger, P. H. *Chem. Commun.* 1100–1102 (2008b).

Hessel, V., Hofmann, C., Lob, P., Lohndorf, J., Lowe, H., and Ziogas, A. *Org. Proc. Res. Dev.* **9**, 479–489 (2005).

Hinchcliffe, A., Hughes, C., Pears, D. A., and Pitts, M. R. *Org. Proc. Res. Dev.* **11**, 477–481 (2007).

Honda, T., Miyazaki, M., Nakamura, H., and Maeda, H. *Adv. Synth. Catal.* **348**, 2163–2171 (2006).

Honda, T., Miyazaki, M., Yamaguchi, Y., Nakamura, H., and Maeda, H. *Lab Chip* **7**, 366–372 (2007).

Hook, B. D.A., Dohle, W., Hirst, P. R., Pickworth, M., Berry, M. B., Booker-Milburn, K. I. *J. Org. Chem.* **70**, 7558–7564 (2005).

Hooper, J., and Watts, P. *J. Labelled Comp. Radiopharm.* **50**, 189–196 (2007).

Kawanami, H., Matsushima, K., Sato, M., and Ikushima, Y. *Angew. Chem. Int. Ed.* **46**, 5129–5132 (2007).

Kenis, P. J.A., Ismagilov, R. F., and Whitesides, G. M. *Science* **285**, 83–85 (1999).

Kikutani, Y., Horiuchi, T., Uchiyama, K., Hisamoto, J., Tokeshi, M., and Kitamori, T. *Lab Chip* **2**, 188–192 (2002a).

Kikutani, Y., Hibara, A., Uchiyama, K., Hisamoto, J., Tokeshi, M., and Kitamori, T. *Lab Chip* **2**, 193–196 (2002b).

Kirschning, A., Altwicker, C., Drager, G., Harders, J., Hoffmann, N., Hoffmann, U., Schonfeld, H., Solodenko, W., and Kunz, U. *Angew. Chem. Int. Ed.* **40**, 3995–3998 (2001).

Kirschning, A., and Gas, J. *Chem. Eur. J.* **9**, 5708–5723 (2003).

Kirschning, A., Solodenko, W., and Mennecke, K. *Chem. Eur. J.* **12**, 5972–5990 (2006).

Kulkarni, A. A. *World Patent*, WO2007087816A1 (2007).

Kulkarni, A. A., Nivangune, N. T., Kalyani, V. S., Joshi, R. A., and Joshi, R. R. *Org. Proc. Res. Dev.* **12**, 995–1000 (2008).

LaPorte, T. L., Hamedi, M., DePue, J. S., Shen, L., Watson, D., and Hsieh, D. *Org. Proc. Res. Dev.* **12**, 956–966 (2008).

Li, X., Wang, H., Inoue, K., Uehara, M., Nakamura, H., Miyazaki, M., Abe, E., and Maeda, H. *Chem. Commun.* 964–965 (2003).

Lundgren, S., Russom, A., Jonsson, C., Stemme, G., Haswell, S. J., Anderson, H., and Moberg, C. *"8th International Conference on Miniaturised Systems for Chemistry and Life Sciences"*, p. 878 (2004).

Mason, B. P., Price, K. E., Steinbacher, J. L., Bogdan, A. R., and McQuade, D. T. *Chem. Rev.* **107**(6), 2300–2318 (2007).

Matsuoka, S., Hibara, A., Ueno, M., and Kitamori, T. *Lab Chip* **6**, 1236–1238 (2006).

Matsushita, Y., Kumada, S., Wakabayashi, K., Sakeda, K., and Ichimura, T. *Chem. Lett.* **35**, 410–411 (2006).

Matsushita, Y., Ohba, N., Kumada, S., Sakeda, K., Suzuki, T., and Ichimura, T. *Chem. Eng. J.* **135S**, S303–S308 (2008).

Matsushita, Y., Ohba, N., Kumada, S., Suzuki, T., and Ichimura, T. *Catal. Commun.* **8**, 2194–2197 (2007).

Mennecke, K., and Kirschning, A. *Synthesis* **20**, 3267–3272 (2008).

Meyer, S., Tietze, D., Rau, S., Schafer, B., and Kreisel, G. *J. Photochem. Photobiol. A: Chem.* **187**, 248–253 (2007).

Miyake, N., and Kitazume, T. *J. Fluorine Chem.* **2003**, 243–246 (2003).

Mukae, H., Maeda, H., Nashihara, S., and Mizuno, K. *Bull. Chem. Soc. Jpn.* **80**, 1157–1161 (2007).

Nagaki, A., Takizawa, E., and Yoshida, J. *J. Am. Chem. Soc.* **131**(5), 1654–1655 (2009).

Nagaki, A., Tomida, Y., and Yoshida, J. *Macromolecules* **41**, 6322–6330 (2008).

Odedra, A., Geyer, K., Gustafsson, T., Gilmour, R., and Seeberger, P. H. *Chem. Commun.* 3025–3027 (2008).

Panke, Schwalbe, T., Stirner, W., Taghavi-Moghadam, S., and Wille, G. *Synthesis* 2827–2830 (2003).

Ponten, F., Gustafsson, T., and Seeberger, P. H. *Chem. Commun.* 1100–1102 (2008).

Ratner, D. M., Murphy, E. R., Jhunjhunwala, M. D., Snyder, A., Jensen, K. F., and Seeberger, P. H. *Chem. Commun.* **5**, 578–580 (2005).

Sakeda, K., Wakabayashi, K., Matsushita, Y., Ichimura, T., Suzuki, T., Wada, T., and Inoue, Y. *J. Photochem. Photobiol. A: Chem.* **192**, 66–171 (2007).

Sato, M., Matsushima, K., Kawanami, H., and Ikuhsima, Y. *Angew. Chem. Int. Ed.* **46**, 6284–6288 (2007).

Schwalbe, T., Autze, V., Hohmann, M., and Stirner, W. *Org. Proc. Res. Dev.* **8**, 440–454 (2004).

Schwalbe, T., Autze, V., and Wille, G. *Chimia* **56**, 636–646 (2002).

Smith, C. D., Baxendale, I. R., Lanners, S., Hayward, J. J., Smith, S. C., and Ley, S. V. *Org. Biomol. Chem.* **5**, 1559–1561 (2007a).

Smith, C. D., Baxendale, I. R., Tranmer, G. K., Baumann, M., Smith, S. C., Lewthwaite, R. A., and Ley, S. V. *Org. Biomol. Chem.* **5**, 1562–1568 (2007b).

Sugimoto, A., Fukuyama, T., Sumino, Y., Takagi, M., and Ryu, I. *Tetrahedron* **65**, 1593–1598 (2008).

Sugimoto, A., Sumino, Y., Takagi, M., Fukuyama, T., and Ryu, I. *Tetrahedron Lett.* **47**, 6197–6200 (2006).

Takasuga, M., Yabuki, Y., and Kato, Y. *J. Chem. Eng. Jpn.* **39**, 772–776 (2006).

Takei, G., Kitamori, T., and Kim, H. B. *Catal. Commun.* **6**, 357–360 (2005).

Tanaka, K., Motomatsu, S., Koyama, K., Tanaka, S., and Fukase, K. *Org. Lett.* **9**, 299–302 (2007).

Thomsen, M. S., Polt, P., and Nidetzky, B. *Chem. Commun.* 2527–2529 (2007).

Tietze, L. F., and Liu, D. *Arkivoc* **viii**, 193–210 (2008).

Uozumi, Y., Yamada, Y. M.A., Beppu, T., Fukuyama, N., Ueno, M., and Kitamori, T. *J. Am. Chem. Soc.* **128**, 15994–15995 (2006).

Ushiogi, Y., Hase, T., Iinuma, Y., Takata, A., and Yoshida, J. *Chem. Commun.* 2947–2949 (2007).

van Meene, E., Moonen, K., Acke, D., and Stevens, C. V. *Arkivoc* **i**, 31–45 (2006).

Wan, Y. S.S., Chau, J. L.H., Gavrilidis, A., and Yeung, K. L. *Chem. Commun.* 878–879 (2002).

Wang, J., Sui, G., Mocharla, V. P., Lin, R. J., Phelps, M. E., Kolb, H. C., and Tseng, H. R. *Angew. Chem. Int. Ed.* **45**, 5276–5281 (2006).

Watts, P., Wiles, C., Haswell, S. J., and Pombo-Villar, E. *Tetrahedron* **58**, 5427–5439 (2002).

Wild, G. P., Wiles, C., Watts, P., and Haswell, S. J. *Tetrahedron* **65**, 1618–1629 (2009).

Wiles, C., and Watts, P. *Chem. Commun.* 443–467 (2007a).

Wiles, C., and Watts, P. *Chem. Commun.* 4928–4930 (2007b).

Wiles, C., and Watts, P. *Eur. J. Org. Chem.* 1655–1671 (2008a).

Wiles, C., and Watts, P. *Org. Proc. Res. Dev.* **12**, 1001–1006 (2008b).

Wiles, C., and Watts, P. *Eur. J. Org. Chem.* 5597–5613 (2008c).

Wiles, C., Watts, P., and Haswell, S. J. *Tetrahedron* **60**, 8421–8427 (2004a).

Wiles, C., Watts, P., Haswell, S. J., and Pombo-Villar, E. *Org. Proc. Res. Dev.* **8**, 28–32 (2004b).

Wiles, C., Watts, P., Haswell, S. J., and Pombo-Villar, E. *Lab Chip* **4**, 171–173 (2004c).

Wiles, C., Watts, P., and Haswell, S. J. *Tetrahedron* **61**, 5209–5217 (2005).

Wiles, C., Watts, P., and Haswell, S. J. *Tetrahedron Lett.* **47**, 5261–5264 (2006).

Wiles, C., Watts, P., and Haswell, S. J. *Chem. Commun.* 966–968 (2007a).

Wiles, C., Watts, P., and Haswell, S. J. *Tetrahedron Lett.* **48**, 7362–7365 (2007b).

Wiles, C., Watts, P., and Haswell, S. J. *Lab Chip* **7**, 322–330 (2007c).

Wiles, C., Watts, P., and Haswell, S. J. *Chem. Commun.* 966–968 (2007d).

Wiles, C., Watts, P., Haswell, S. J., and Pombo-Villar, E. *Tetrahedron* **59**, 2886–2887 (2003).

Wootton, R. C.R., Fortt, R., and de Mello, A. J. *Org. Proc. Res. Dev.* **6**, 187–189 (2002).

Yoon, S. K., Choban, E. R., Kane, C., Tzedakis, T., Kenis, P. J.A. *J. Am. Chem. Soc.* **127**, 10466–10467 (2005).

Yoshida, J. "Flash Chemistry: Fast Organic Synthesis in Microsystems". Wiley, London (2008).

Zhang, X., Stefanick, S., and Villani, F. J. *Org. Proc. Res. Dev.* **8**, 455–460 (2004).

CHAPTER 4

Microfluidic Reactors for Nanomaterial Synthesis

S. Krishnadasan, A. Yashina, A.J. deMello and **J.C. deMello**[*]

Abstract

The difficulty of preparing nanomaterials in a controlled, reproducible manner is a key obstacle to the proper exploitation of many nanoscale phenomena. An automated chemical reactor capable of producing (on demand and at the point of need) high-quality nanomaterials, with optimized physicochemical properties, would find numerous applications in nanoscale science and technology, especially in the areas of photonics, optoelectronics, bio-analysis, and targeted drug-delivery. In addition such a device would find immediate and important applications in toxicology, where it is essential to characterize the physiological effects of nanoparticles not only in terms of chemical composition but also in terms of size, shape, and surface functionalization. In this chapter, we describe recent advances in the development of microfluidic reactors for

Department of Chemistry, Imperial College London, South Kensington, London SW7 2AY, UK.

[*] Corresponding author:
E-mail address: j.demello@imperial.ac.uk

Advances in Chemical Engineering, Volume 38
ISSN: 0065-2377, DOI 10.1016/S0065-2377(10)38004-5

© 2010 Elsevier Inc.
All rights reserved.

controlled nanoparticle synthesis and, more specifically, work in our group aimed at developing just such an automated reactor.

1. INTRODUCTION

The discovery in recent years of novel properties, processes, and phenomena at the nanoscale has created revolutionary opportunities for the creation of novel materials and devices with superior chemical, physical, and/ or biological characteristics (Ozin and Arsenault, 2005). Nanocrystalline materials are of particular interest in this regard owing to their tunable physico-chemical properties and their potential use as functional elements for biological sensing, optoelectronics, fiber-optic communications, and lasers (Alivisatos, 1996; Euliss et al., 2006; Green, 2004; Han et al., 2007; Kawazoe et al., 2006; Klostranec and Chan, 2006; Matsui, 2005; Medina et al., 2007; Rhyner et al., 2006). The characteristics of nanocrystals are strongly influenced by their physical dimensions, and there is consequently considerable interest in processing routes that yield nanoparticles of well-defined size and shape (Donega et al., 2003).

There are two main routes to nanoparticle formation: (1) "top-down" and (2) "bottom-up" approaches. In top-down routes, nanometer-sized structures are engineered from bulk materials using a combination of lithography, micromachining methods, and etching (Xia et al., 2007). Significantly, the creation of sub-100-nm structures requires lithographic techniques beyond the optical domain, such as electron beam and X-ray lithography. Such approaches are technically challenging and although reproducible do not readily lend themselves to large-scale production. A bottom-up approach on the other hand involves the chemical growth of particles on an atom-by-atom or molecule-by-molecule basis until the desired particle size and shape are achieved (Malik et al., 2005). This growth process occurs spontaneously in super-saturated solutions and has been successfully used to create high-quality spherical, cubic, tubular, and tetrahedral crystallites in kilogram quantities and above (Masala and Seshadri, 2004; Milliron et al., 2004; Park et al., 2004). The bottom-up approach, which may be carried out on the laboratory bench using standard techniques in synthetic chemistry, has attracted considerable interest owing to its versatility and ease of use, and it is by far the most practical and prevalent means of producing large quantities of nanomaterials. In recent years, a variety of sophisticated chemical strategies have been reported for producing near defect-free nanoparticles of consistent size, shape, and chemical composition. Their implementation, however, remains a complex and difficult undertaking that requires a combination of skill (and often luck), intuition, and extensive experimentation to obtain well-defined nanoparticles with tightly specified properties (Donega et al.,

2003; Dushkin et al., 2000; Rao et al., 2006; Yordanov et al., 2006). Indeed the formidable difficulty of preparing high-quality nanoparticles in a controlled and reproducible manner is recognized to be the foremost obstacle to the full exploitation of many nanoscale phenomena. In this chapter, we consider the utility of microfluidic systems in providing a controlled environment in which to synthesize nanomaterials and also the feasibility of automating the synthesis procedure, with a view to creating "blackbox" chemical reactors that in response to appropriate instructions—and without any human intervention—can produce high-quality nanomaterials with optimized physicochemical properties. The provision of such automated tools would significantly advance the field of nanoscience, allowing high-quality nanomaterials to be produced on-demand and at the point of need for a host of applications in nanotechnology.

2. MICROFLUIDIC ROUTES TO THE SYNTHESIS OF NANOPARTICLES

The approach described later on in this chapter builds upon a report in 2002, in which we proposed microfluidic reactors as favourable systems for nanoparticle synthesis, and showed that nanocrystalline cadmium sulfide prepared in such reactors exhibited improved monodispersity compared with particles prepared in conventional bulk-scale vessels (Edel et al., 2002).

In simple terms nanoparticles are formed via a two-stage process in which an initial nucleation stage (in which seed particles spontaneously precipitate from solution) is followed by a more gradual growth phase in which diffusion of solutes from the solution to the seed surface proceeds until the final particle size is attained. This classical model was first proposed by LaMer and Dinegar to explain the mechanism of formation of sulfur colloids (La Mer and Denegar, 1950) and can be represented by a simple cartoon of the kind shown in Figure 1 which shows the variation in solute concentration as a function of time. The solute is formed by a chemical reaction (e.g., hydrolysis of a metal alkoxide, hydration of metal ions, and decomposition of organic compounds). As the reaction proceeds, the solute concentration will at some point exceed a "supersaturation" concentration, eventually reaching a "critical" concentration at which point nucleation occurs (ideally) in a short burst. This nucleation process—and the ensuing growth of these nuclei—lowers the solute concentration to a value below the critical nucleation concentration, thereby preventing further nucleation, but still allowing particle growth. The formed particles then grow at a rate that merely consumes all further solutes that are generated by the chemical reaction. The process of particle growth will lower the overall free energy of the system so, in the absence of any other competing processes, growth will

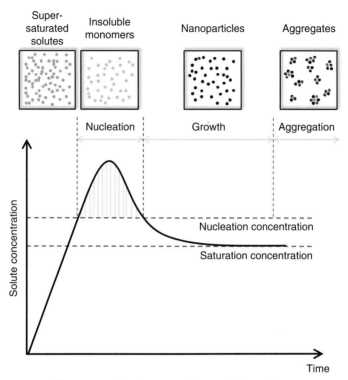

Figure 1 Cartoon illustration of nucleation and growth during the preparation of monodisperse nanoparticles.

continue until all of the solute has been consumed. Additionally, aggregation of individual particles also lowers the free energy of the system so the particles will tend to coalesce over time and precipitate out of solution. Unfortunately, in most cases, nucleation and growth occur concurrently, and thus the final particle population will therefore exhibit a broad (and undesirable) size distribution. Accordingly, to obtain monodisperse particles, it is necessary to set up conditions in which all nucleation takes place over a short period of time with additional material being supplied so gradually that it find its way to the nuclei without the solute concentration reaching a level at which further nucleation can take place. In practical terms it is crucial that all nuclei should form and grow in an identical environment with state functions (such as temperature, pressure, and concentration) assuming constant values throughout the reaction volume. In conventional syntheses within bulk reactors (where turbulent mixing is used to ensure rapid combination of reagents) significant variations in physical conditions across the reaction chamber are typical, thus generating wide particle-size distributions. In the current context, microfluidic systems—which allow for rapid and controlled thermal and mass transfer—are an ideal format for nanoparticle production. They are able to control the temperature or temperature gradient along a flow profile and can rapidly heat or cool the reagent mixture. They can efficiently mix reagents on sub-millisecond time scales and allow reagents to be added in a flexible and controlled manner.

In subsequent studies by ourselves and others, a variety of microfluidic architectures have been used to prepare metal and compound semiconductors nanoparticles, including CdS, CdSe, TiO$_2$, Ag, Au, and Co (Boleininger et al., 2006; Chan et al., 2003, 2005; Cottam et al., 2007; DeMello and DeMello, 2004; Khan et al., 2004; Kohler et al., 2005; Krishnadasan et al., 2004; Millman et al., 2005; Shalom et al., 2007; Shestopalov et al., 2004; Song et al., 2006; Takagi et al., 2004; Wagner and Kohler, 2005; Wang et al., 2004, 2005; Xue et al., 2004; Yen et al., 2003, 2005). In simple terms, microfluidic devices manipulate and process sub-microliter volumes of liquid in enclosed channels that are typically no more than a few hundred microns in diameter (see Figure 2). Importantly for chemical synthesis, they have a number of advantageous features, including precise control over reaction conditions, rapid heating, cooling and mixing of fluid streams, and the ability to combine multiple chemical processes into a single, integrated device (deMello, 2006). Over the years, numerous functional elements have been successfully integrated into microfluidic devices (including filters, heaters, mixers, valves, actuators, electrophoretic separators, distillers, and various kinds of transducer), and it is now possible to perform virtually every conceivable chemical process using appropriate combinations of these elements. This has given rise to the concept of a lab-on-a-chip, a self-contained integrated device that is capable of performing all steps in a complete chemical synthesis or analysis without any additional equipment or instrumentation.

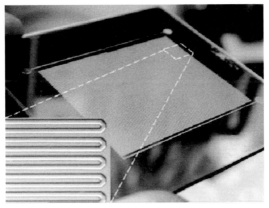

Figure 2 A simple microfluidic device fabricated in glass using wet lithography. The typical channel dimensions in a microfluidic device are 5–100 μm.

In the early days, microfluidic devices were used mainly for chemical analysis due to their intrinsic ability to handle small sample volumes and the significant performance enhancements that typically arise when analytical techniques are transferred to the microscale (Manz et al., 1992; Vilkner et al., 2004). However, in recent years, there has been increasing interest in using them for chemical synthesis where the precise control they provide over reaction conditions offers potential improvements in product yield, purity, and quality (deMello, 2006). Microfluidic devices can be fabricated using a variety of substrate materials, including glass, silicon, and polymers such as poly(methyl-methacrylate) and poly-(dimethylsiloxane) (Dumais et al., 2006; Iliescu, 2006). Of these materials, glass and quartz are generally preferred for nanoparticle synthesis due to their transparency in the visible region of the spectrum and tolerance of elevated reaction temperatures. Glass devices are normally prepared using wet chemical etching (see reference Iliescu, 2006), but other techniques such as direct laser-etching procedures have also been used successfully. Devices of considerable sophistication have been reported in the literature, as shown, for example, in Figure 3, but the benefits of miniaturization can also be exploited in surprisingly simple chip architectures as will be seen below.

Figure 3 A sophisticated microfluidic device used for DNA processing. Numerous functional elements have been adapted for use with microfluidics, and it is now possible to perform virtually any combination of chemical processes in a microfluidic device.

Figure 4   A simple y-shaped microfluidic device with two inlet channels and a single outlet channel. The device can be used to mix and react two reagents A and B. By varying the relative volumetric rates (F_A and F_B) at which the reagents are injected, it is possible to vary the composition of the mixture in the outlet channel. By varying the total flow rate ($F = F_A + F_B$), it is possible to control the time the reagents spend in the reaction channel.

Microfluidic reactors comprise one or more inlet ports into which reagents are typically injected using precision syringe pumps and capillary connectors, a network of channels in which various chemical processes are carried out, and one or more outlet ports where products and waste materials are extracted. Figure 4 for instance shows a simple y-shaped chemical reactor with two inlet channels and a single outlet channel that can be used to mix and react two reagents A and B. The composition of the resultant mixture and the residence time inside the microfluidic chip can be controlled by varying the volumetric flow rates of the two reagent solutions, F_A and F_B. Hence, if the volume of the outlet channel is V and the molar concentrations of the two reagent solutions are [A] and [B], then the residence time (τ) in the reaction outlet channel is given by

$$\tau = \frac{V}{F_A + F_B} \tag{1}$$

and the molar ratio (R) of A to B in the final mixture is equal to.

$$R = \frac{F_A}{F_B} \times \frac{[A]}{[B]} \qquad (2)$$

In a typical scenario, the volume of the outlet channel might be $10\,\mu l$ and the individual injection rates of the two reagent solutions might each be of order $100\,\mu l\,min^{-1}$. Using modern syringe pumps, it is straightforward to control each of the injection rates to within $0.1\,\mu l\,min^{-1}$, and so in these circumstances τ and R can each be controlled to within 0.1%, which far exceeds the precision achievable in conventional macroscale reactors. In the case of thermally initiated reactions, the y-shaped reactor can be placed on a hot-plate with high spatial uniformity. In this way, the entire reaction volume can be held at a uniform temperature that can be controlled to within a fraction of a degree, which similarly exceeds the typical temperature uniformity achievable in a conventional macroscale reactor.

An additional advantage of using microfluidic devices, which we do not have the space to discuss in detail here, is the absence of turbulence (Koo and Kleinstreuer, 2003). In the context of nanoparticle synthesis, turbulence gives rise to unpredictable variations in physical conditions inside the reactor that can influence the nature of the chemical product and in particular affect the size, shape, and chemical composition. In microfluidic devices, turbulence is suppressed (due to the dominance of viscous over inertial forces) and fluid streams mix by diffusion only. This leads to a more reproducible reaction environment that may in principle allow for improved size and shape control.

These improvements in control do not come without cost since the transferral of chemical reactions from the macroscale to the microscale frequently requires compromises to be made in terms of reagent selection and reaction conditions. The solvents used for microfluidic syntheses, for instance, should ideally be low-viscosity liquids at room temperature $(mp \ll 20\,°C)$ as this conveniently avoids the need to heat the syringe pumps or to thermally lag the capillaries that connect the syringe pumps to the inlets of the microfluidic device. And, in the case of standard lyothermal syntheses, the solvents should also have sufficiently high boiling points to permit thermal initiation of the chemical reactions $(bp \gg 250\,°C)$. The former requirement precludes the use of many of the standard solvents used in conventional macroscale synthesis routes (such as long-chain alkylamines) as these tend to be waxy materials with high melting points. Unfortunately, solvents with low melting points frequently have relatively low boiling points and so tend to be unsuitable for many lyothermal syntheses. The choice of solvent system for microfluidic-based syntheses is therefore somewhat restricted compared with macroscale syntheses. In addition, it is generally not possible to use the extremely high reaction temperatures favored for many macroscale lyothermal syntheses ($T > 300\,°C$) since above about $280\,°C$ glass chips have a tendency to fracture and the epoxy-based glues used to attach the capillaries to the microfluidic chips tend to decompose. The reactions must therefore be carried out at lower temperatures, which risks increasing the number of defects in the resultant nanocrystals. These issues, although mundane in nature, mean it is rarely possible to implement existing synthesis routes directly on a microfluidic chip without making at least some adaptations to the solvent, reagent mixtures, and/or reaction temperature.

In this chapter, we will focus on the automated preparation of CdSe quantum dots (which have been extensively studied and characterized in the scientific literature), but the principles and techniques we describe are general ones that may be applied to a variety of nanomaterials. Indeed, the following paragraphs outline some selected studies which have reported the use of microfluidic reactors for the synthesis of compound semiconductor nanomaterials.

Nakamura and coworkers have reported the synthesis of CdSe nanoparticles in simple glass capillaries (Nakamura et al., 2002). The authors utilized the ability to control both temperature and the mixing environment to rapidly synthesize CdSe particles from pre-mixed trioctylphosphine: selenium and cadmium acetate. Using such an approach, they were able to continuously produce CdSe nanoparticles at temperatures between 230 and $300\,°C$ in 7–$150\,s$. The size and quality of the obtained nanoparticles were assessed using absorbance spectroscopy and residence time distributions were reduced by introducing $500\,nl$ nitrogen bubbles into the flow at defined intervals. Subsequently, the same group advanced the complexity of the synthesis process nanoparticles by generating CdSe/ZnS core/shell nanoparticles on the microscale (Wang et al., 2004). Specifically, preprepared reagents (CdSe, ZnS) were allowed to react within a chip-based microfluidic reactor, and two separate oil baths were used to maintain defined temperatures for the nucleation and growth of CdSe and ZnS, respectively (Figure 5a). CdSe nanoparticles with different ZnS shell thicknesses could be generated by simply varying the flow rate and capillary length. Luminescence spectra of uncoated CdSe and ZnS-coated particles are shown in Figure 5B, demonstrating the utility of the process. More recently the authors have refined their approach to demonstrate the synthesis of ZnS/CdSe/ZnS quantum-dot quantum-well (QDQW) structures (Uehara et al., 2009). In essence QDQWs are analogous to planar quantum-well devices where the narrow band gap inner shell acts as a quantum well embedded in a wide-gap material. By altering the size of the core and the thickness of both inner and outer shells, the electronic properties of the QDQW may be controlled. In many respects continuous flow microfluidic systems are ideal tools for synthesizing QDQWs since both

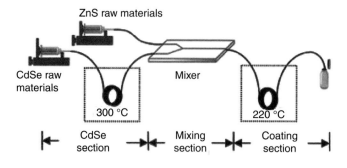

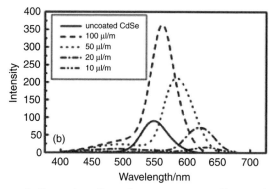

Figure 5 A schematic illustration of a multistep continuous-flow synthesis system for making ZnS-coated CdSe composite particles. Luminescence spectra of uncoated CdSe from CdSe portion and coated particles from the outlet at different flow rates. Labels denote the flow volume of the two syringes for the CdSe and ZnS raw feedstock. Uncoated CdSe nanoparticles were obtained using a flow rate of 100 µl min^{-1} images reproduced, with permission, from Wang et al., 2004).

the heating time and temperature can be controlled precisely. To demonstrate efficacy, the authors mixed a ZnS colloid solution with a core CdSe population in a continuous flow at 240 °C. The formed ZnS/CdSe (core/shell) solution was subsequently mixed with ZnS at 150 °C to produce the ZnS/CdSe/ZnS (core/shell/shell) multilayer composite nanoparticles. Such a high temperature synthesis is favorable since it leads to a high-quality crystalline coating, but requires careful control to avoid Ostwald ripening of cores that would lead to particle polydispersity. The authors importantly showed that the resulting ZnS/CdSe/ZnS QDQWs emit blue fluorescence with a fluorescence quantum efficiency as high as 50%. Moreover, photoluminescence could be tuned with ease by varying the volumetric flow rate during the CdSe deposition process.

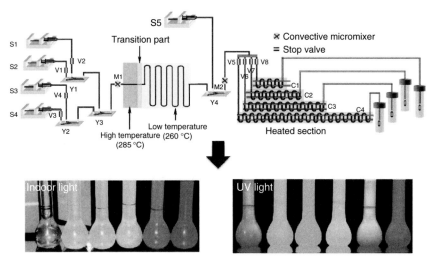

Figure 6   Schematic of an integrated microreactor for the continuous synthesis of CdSe/ZnS and CdS/ZnS nanoparticles (S1-syringe pump with Se precursor, S2-syringe pump with S precursor, S3-syringe pump with Cd-OA-OLA, S4-syringe pump with Cd-OA-OLA-TOPO, Y—Y conjunction, M-micromixer, V-stop valve, C-channel;images reproduced, with permission, from Yang et al., 2009).

More recently, Yang et al. have extended this idea and demonstrated the ability to perform multistep reactions in a sequential manner by producing ZnS-capped CdSe and CdS nanocrystals in a microfluidic device (Yang et al., 2009). This system involved distinct temperature zones for nucleation and growth of the core and a serpentine reaction channel for coating the ZnS shell (Figure 6). Such a system provides for both efficient mixing and uniform residence times for the precursors at high volumetric flow rates. The authors demonstrated the continuous synthesis of strongly emitting CdS/ZnS and CdSe/ZnS nanocrystals through real-time adjustment of precursor and capping inputs. Importantly, luminescence quantum efficiencies ranged from 41% for blue emitting particles and 78% for yellow particles.

A number of elegant studies over the past few years have also addressed the need to minimize particle size distributions through the use of segmented flow microfluidic systems. Such an approach removes the possibility of particle deposition on channels and eliminates the problems of residence time distributions that occur in single phase systems (where drag at the channel walls sets up a velocity distribution inside the channel). For example, Shestopalov et al. reported the two-step synthesis of colloidal CdS and CdS/CdSe core–shell nanoparticles in a droplet-based microreactor (Shestopalov et al., 2004). In addition, Chan et al. have described the use of microfluidic droplet reactors for the high-temperature synthesis of CdSe nanoparticles (Chan et al., 2005). In this study, Cd/Se precursor solutions were made to form stable nanoliter-sized droplets in a perfluorinated polyether continuous phase. The encapsulated reagents were reacted when heated to 290 °C to yield nanoparticles 3.4 nm in diameter. In a similar study, Jensen et al. used gas–liquid (rather than liquid–liquid) segmented-flow reactors incorporating distinct temperature zones for the synthesis of high-quality CdSe quantum dots (Figure 7) (Yen et al., 2005). More recently, Hung et al. synthesized CdS nanoparticles in droplets by the passive fusion of the droplets containing different reagents ($Cd(NO_3)_2$ and NaS) (Hung et al., 2006). Controlled passive fusion was achieved using a dilating channel geometry and controlled liquid-phase flow. Particles obtained using this approach were generally smaller than those obtained by bulk and ranged between 8.2 and 4.2 nm in diameter. Significantly, all of the above studies leveraged efficient mixing and reduced residence-time distributions to engender improvements in both yield and size distribution. As an aside, it is noted that higher order nanostructures (inaccessible via conventional routes) can be synthesized using static microdroplet reactors. For example, Millman et al. reported the synthesis of anisotropic particles using nanoliter-sized droplets which are made to float on the surface of a perfluorinated oil (Millman et al., 2005). Since such floating droplets can be controlled by electrical fields, droplets containing suspensions of polymers and nanoparticles can be persuaded to form complex particle structures. Indeed, "striped" multilayer particles were generated from ternary mixtures of gold, fluorescent latex, and silica particles; and core–shell particles could be synthesized by the encapsulation of droplets of aqueous suspensions inside polymer shells.

Interestingly, the majority of compound semiconductor research to date has focused on II–VI systems such as CdSe due to the relative simplicity of the synthetic routes. III–V materials, whilst potentially less toxic, have presented a stiffer synthetic challenge. In large part, the strong covalency of the constituent atoms makes it hard to devise labile precursors, which leads to synthesized particles with permanent crystal defects. Very recently, Nightingale and deMello reported the first synthesis of III–V (InP) nanoparticles using a microfluidic reactor (Nightingale and de Mello, 2009). Using a simple microfluidic mixer (maintained at a temperature in excess of 200 °C) and product detection using a 355 nm diode-pumped Nd:YAG laser and a fiber-optic-coupled charge-coupled device (CCD) spectrometer, the authors verified that microfluidic reactors can be readily applied to III–V materials, yielding particles that are of comparable quality to those obtained using bulk methods.

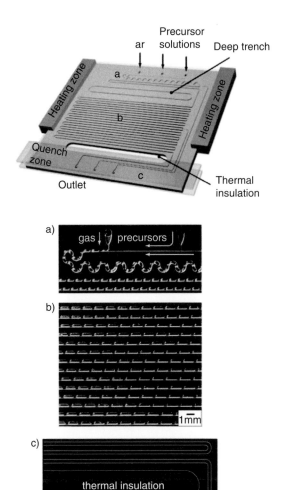

Figure 7 A microfluidic reactor for CdSe nanoparticle production. (A), The reactor provides for rapid precursor mixing (sector A), particle growth (sector B), and reaction quenching (sector C). The reactor accommodates a 1-m-long reaction channel and two side channels for collecting reaction aliquots. A halo etch region allows localization of temperature zones for reaction (>260 °C) and quenching (<70 °C). Precursor solutions are delivered into the heated section separately, and an argon gas stream generates a segmented gas–liquid flow. Recirculation within the liquid slugs rapidly mixes reagents and initiates the reaction. The reaction is stopped when the fluids enter the cooled outlet region. Photographs of heated inlets (B) and main channel section (C). Red segments show the reaction solution; dark segments define Ar gas; $T = 260$ °C; gas flow rate $= 60\,\mu l\,min^{-1}$; liquid flow rate $= 30\,\mu l\,min^{-1}$; images reproduced, with permission, from Yen et al. (2005).

3. THE AUTOMATED PRODUCTION OF CdSe NANOPARTICLES

The synthetic route we use in this work is a simple adaptation of a method by Peng et al., in which CdO and elemental Se are reacted together at high temperature in the presence of oleic acid to form CdSe nanocrystals (Peng and Peng, 2001). In brief, a precursor Se solution is prepared by combining 30 mg of Se with 10 ml of 1-octadecene and 0.4 ml of trioctylphosphine and warming over a hot plate. A Cd precursor solution is prepared by combining 13 mg of CdO and 0.6 ml of oleic acid in 10 ml of 1-octadecene and heating at 180 °C until clear. The reaction is performed in a glass y-shaped microfluidic chip with channels of width 330 μm and depth 160 μm. The reaction channel is 40 cm long and arranged in a serpentine architecture for compactness. The chip is placed on a stabilized hot plate with high spatial uniformity. The reaction can be performed at temperatures in the range 160–255 °C. Two syringe pumps are used to inject the precursor solutions into the inlet channels at rates up to 40 μl min^{-1}. The solutions mix rapidly at the point of confluence, and nucleation and growth of the CdSe nanoparticles occurs along the reaction channel. The emergent particles can then be monitored at ambient temperature at an observation zone downstream of the reaction zone. Here, the particles are excited using a 355-nm solid-state laser excitation source and emission is detected using a fiber-optic CCD spectrometer. A fraction of the incident laser light is redirected to a Si photodiode using a beam-splitter, allowing the emission spectra to be corrected for variations in the laser intensity. The set up is shown in Figure 8.

Figure 9a shows a TEM image of a typical CdSe nanoparticle prepared using the above synthesis. The emission spectrum of the nanoparticles obtained at an illustrative reaction temperature of 220 °C and equal

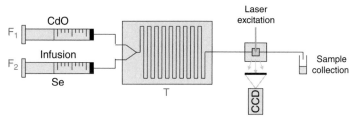

Figure 8 A schematic of the reactor used to synthesize the nanoparticles described in this chapter. Cd and Se precursor solutions are stored in two separate syringes and injected at flow rates **F₁** and **F₂** into the two inlets of a y-shaped microfluidic device. The microfluidic device rests on a hot plate of variable temperature **T**. The reagent streams meet at the point of confluence and nucleation, and growth of the particles occurs as they pass along the outlet channel. The emission spectra of the particles so produced are monitored prior to collection at a detection-zone downstream of the chip using a 355-nm Nd:YAG laser as an excitation source and a fiber-optic-coupled CCD spectrometer.

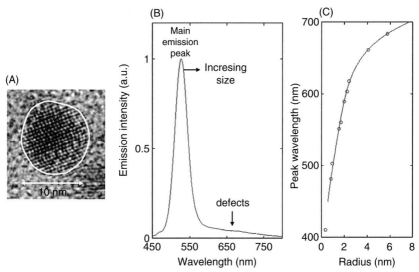

Figure 9 (A) Typical TEM image of a CdSe nanoparticle synthesized by the direct reaction of Se and CdO (see main text). (B) Typical emission spectrum of a CdSe nanoparticle. The spectrum exhibits two main features: (1) a sharp Gaussian-shaped peak due to band-edge emission and (2) a broad feature at lower energies due to defect emission. (C) The size-dependence of the peak wavelength for CdSe nanoparticles, determined using data provided in Murray et al. (1993).

CdO and Se flow rates of $9 \, \mu l \, min^{-1}$ is shown in Figure 9b. The spectrum comprises a strong band-edge emission peak at 525 nm with full-width half-maximum 35 nm, and a broad weaker peak at 660 nm due to emission from crystal defects at the surface of the nanoparticles. Murray et al. have previously reported emission data for a series of size monodisperse nanoparticles, and it can be seen from their data (reproduced in Figure 9c) that the wavelength of the band-edge emission peak shifts progressively to longer wavelengths with increasing particle size, consistent with reduced quantum confinement effects.

The power of the microfluidic approach is apparent in Figures 10 and 11 where we consider the effects of varying T and τ. Increasing the temperature and extending the reaction time have similar effects on the band-edge emission, leading in both cases to a red-shift and enhancement in the intensity. This is consistent with the formation at higher temperatures and longer reaction times of larger particles, in which excitons are less tightly constrained and so less susceptible to trapping at surface defects (Pradhan et al., 2003). In both cases, a remarkable degree of control is achieved over the nanoparticle properties—with, for example, the peak wavelength being tunable to within a fraction of a nanometer—indicating the power of the microfluidic approach.

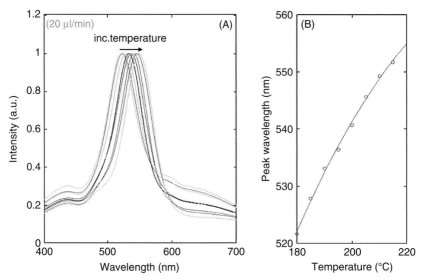

Figure 10 (A) Temperature dependence of the emission spectra of CdSe nanoparticles prepared at a fixed flow rate of 20 μl min⁻¹ using the microfluidic system shown in Figure 8; (B) Temperature dependence of the peak wavelength, determined from the data in (A).

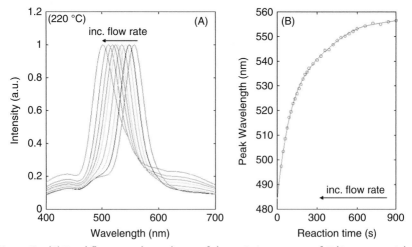

Figure 11 (A) Total flow-rate dependence of the emission spectra of CdSe nanoparticles prepared at a fixed temperature of 220 °C using the microfluidic system shown in Figure 8; (B) Total flow-rate dependence of the peak wavelength, determined from the data in (A).

4. THE AUTOMATED PRODUCTION OF NANOPARTICLES

As noted in the introduction, a major aim of the current research is the development of "black-box" automated reactors that can produce particles with desired physicochemical properties on demand and *without* any user intervention. In operation, an ideal reactor would behave in the manner of Figure 12. The user would first specify the required particle properties. The reactor would then evaluate multiple reaction conditions until it eventually identified an appropriate set of reaction conditions that yield particles with the specified properties, and it would then continue to produce particles with exactly these properties until instructed to stop. There are three essential parts to any automated system—(1) physical machinery to perform the process at hand, (2) online detectors for monitoring the output of the process, and (3) decision-making software that repeatedly updates the process parameters until a product with the desired properties is obtained. The effectiveness of the automation procedure is critically dependent on the performance of these three subsystems, each of which must satisfy a number of key criteria: the machinery should provide precise reproducible control of the physical process and should carry out the individual process steps as rapidly as possible to enable fast screening; the online detectors should provide real-time low-noise information about the end product; and the decision-making software should search for the optimal conditions in a way that is both parsimonious in terms of experimental measurements (in order to ensure a fast time-to-solution) and tolerant of noise in the experimental system.

The microfluidic system described above has two key features that make it especially amenable to creating such automated reactors: firstly, the reaction conditions can be precisely and rapidly manipulated (which in turn means that it is possible to finely tune the physical properties of the reaction product), and secondly, the inline spectrometer provides immediate real-time information about the product. The only remaining element

Figure 12   Schematic illustrating the desired behavior of an automated chemical reactor. The user enters the desired particle properties, the "black-box reactor then evaluates multiple reaction conditions until it identifies an appropriate set that yield particles with the desired properties; the reactor then continues to produce particles with these properties until instructed to stop.

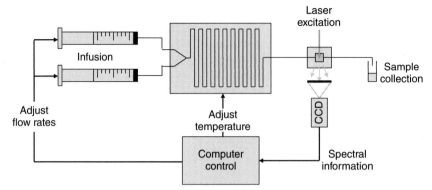

Figure 13 Schematic of an automated system for producing nanoparticles with desired properties. The set up is an adaptation of the system shown in Figure 8. The emission spectra of the emergent nanoparticles recorded by the CCD are passed to an intelligent control algorithm that repeatedly updates the reaction temperature and the injection rates of the two reagents until particles with the desired properties are obtained.

needed to create a fully automated reactor is the addition of a suitable control algorithm that is able to repeatedly update the reaction conditions until particles with the desired properties are obtained (see Figure 13).

The intention of the remaining part of this chapter is to explain in simple terms how the process of nanoparticle synthesis can be automated, and we will not dwell excessively on mathematical details. Instead, to illustrate our general approach, we will start by considering a very simple example in which we will design an automated system capable of producing particles that emit at a specified wavelength. We will then show how this approach can be generalized to enable the automated production of particles whose emission characteristics are specified in more complex ways. The key to automating the process of nanoparticle synthesis is to use a so-called *utility function* that reduces all of the known information about the particles to a single figure of merit that characterizes the particle quality. The utility function is usually defined in such a way that the figure of merit decreases steadily (from a large positive number) to zero as the measured properties get progressively closer to the desired properties. For example, let us suppose that one wishes to produce particles that emit at a certain target wavelength λ_t and that for the current reaction conditions the measured wavelength is λ_c. In this case, a sensible utility function $u(\lambda_c)$ would be

$$u(\lambda_c) = (\lambda_c - \lambda_t)^2 \tag{3}$$

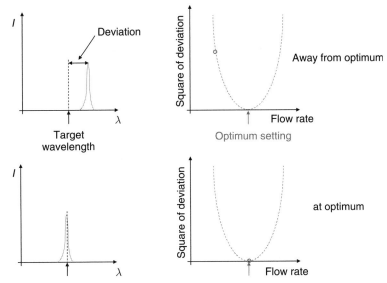

Figure 14 Schematic illustrating the use of a "utility function" to optimize the emission wavelength by varying the total flow rate of the reagents. The utility function generates a figure of merit that equals the square of the deviation of the current wavelength from the target wavelength. The square of the deviation is large and positive when the current wavelength is far from the target and decreases to zero as the target wavelength is approached. The optimization routine adjusts the flow rate in an effort to minimize the figure of merit and, in so doing, indirectly finds the reaction conditions that yield particles with the desired emission properties. The same approach can be readily extended to multiple reaction variables and the simultaneous optimization of multiple attributes (e.g., peak wavelength, peak intensity, and line-width).

which has a value of zero when the current wavelength equals the target wavelength and becomes progressively larger in size as the current wavelength deviates (in either direction) from the target.[1] The minimization routine tries to minimize the figure of merit and, in so doing, indirectly finds the reaction conditions that yield particles that emit with the target wavelength as shown in Figure 14. The precise manner in which the routine goes about finding the minimum is clearly crucial to the success of the optimization, and a number of issues need to be taken into account in choosing or designing a suitable routine:

1. The detailed mechanisms of nanoparticle formation (nucleation, growth, aggregation, and ripening) are understood only in qualitative detail so there are no reliable process models available

---

[1] Note, an alternative figure of merit would be the absolute deviation from the target wavelength. This choice of utility function, however, would exhibit an abrupt change of slope at the optimum which is liable to cause numerical difficulties for the control algorithm. The slope of the proposed parabolic utility function varies smoothly about the optimum and hence avoids these difficulties.

to guide the automation. This differs from the usual situation in chemical engineering where mathematical models of the chemical plant are normally available to supplement measured data. In the case of nanoparticle synthesis, the only information available to the algorithm is the measured data. The utility function in effect behaves like a black box that converts the process conditions to a final figure of merit in a manner that is entirely unknown to the minimization routine.

2. The same nominal reaction conditions may on different occasions give rise to slightly different products due to mechanical or chemical imperfections in the reactor. The utility function is therefore said to be noisy. The vast majority of minimization routines are designed for handling noise-free mathematical functions and only a very small subset of algorithms can cope with appreciable noise.

3. Owing to the slow nature of nanoparticle formation, there is typically a delay of several minutes between setting new reaction conditions and the figure of merit reaching a new stable value. The utility function is said to be expensive to evaluate and, from a practical perspective, this means only a small number of reaction conditions can be tested during the search for the optimum process conditions. This restricts the choice of algorithm severely since conventional minimization routines often require thousands of measurements to find the optimum conditions. If each measurement takes on average 5 min, one thousand measurements would correspond to a full week of searching, which is clearly excessive for many applications. Ideally, the control algorithm should be able to find the optimum reaction conditions in one hundred measurements or less.

4. The reaction conditions are constrained. In other words, there is usually a strict upper and lower limit for each reaction parameter. In the case of the synthesis described above, for example, the lower temperature is set by the need to provide sufficient thermal energy to initiate the reaction and the upper temperature by the need to remain below the decomposition temperature of the glue (see Section 2). The lower and upper limits on the total flow rate meanwhile are determined, respectively, by the maximum length of time one is prepared to allow for a single reaction and the minimum reaction time needed to produce crystals of nanometer dimensions. In this work, we select minimum and maximum total flow rates of 2 and $40 \, \mu l \, min^{-1}$ which, for the typical chip volumes we use ($\sim$16.6 μl), correspond to average residence times of about 500 and 25 s, respectively.

5. The majority of minimization routines are designed for unconstrained optimization, in which the control algorithm is free to select any parameters it wishes. Only a minority can handle constrained optimization.

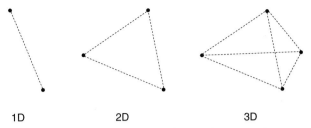

1D   2D   3D

Figure 15 Schematic illustrating the concept of a simplex. A simplex is a geometric shape formed from $N+1$ vertices in an N-dimensional space. Hence, for one, two- and three-dimensional spaces, the simplex points comprise the vertices of a line, triangle, and tetrahedron, respectively. The simplices can move through their respective spaces by undergoing repeated reflections and/or changes of shape (see main text).

There are very few minimization routines that satisfy all of these conditions (i.e. that can perform constrained optimisation of expensive noisy black-box functions), and it is only in the past few years that effective algorithms have started to emerge. One promising technique, developed originally for noise-free optimization, is the simplex method (Kolda et al., 2003). A simplex is a geometric shape formed from $N+1$ distinct points (vertices) in an N-dimensional space. This is shown in Figure 15a–c for one-, two-, and three-dimensional spaces, where it can be seen that the simplex points comprise the vertices of a line, triangle, and tetrahedron, respectively. In the most basic form of simplex optimization, the value of the function is evaluated at each of the vertices and the simplex is repeatedly reflected away from the worst vertex. This is illustrated in Figure 16 for the illustrative two-dimensional function $f(x,y) = x^2 - 4x + y_2 - y - xy$. With each successive reflection, the simplex moves progressively closer to the minimum at (3,2). The simplex ceases to make progress at the 28th iteration (T28). The reason is straightforward. Vertex A has a larger function value than B and C and is therefore reflected to D. The new vertex D also has a larger value than B and C and at the next reflection is reflected back to A. The simplex therefore oscillates between ABC and BCD and makes no further progress toward the minimum. The premature "stalling" of the search is a limitation of the simple "reflective-simplex" approach, and the procedure can be improved substantially by using an "adaptive-simplex" that changes its size and shape according to the shape of the local terrain. This has two key advantages as illustrated in Figure 17: (1) the simplex can expand in size when moving through flat uninteresting terrain, which improves the efficiency of optimization, and (2) once it is close to the location of the optimum, it can start to contract, which improves the accuracy of the final solution. For the situation considered in Figures 16 and 17, after just 10 iterations the adaptive simplex is closer to the minimum than the reflective simplex was after 28.

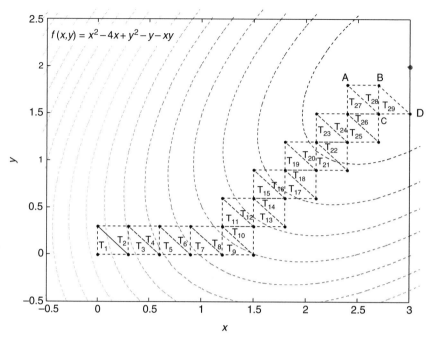

Figure 16 Schematic illustrating the concept of simplex optimization using a simple reflective algorithm, in which the simplex vertex with the largest function value is repeatedly reflected through the line joining the other two vertices. The simplex moves progressively closer to the minimum at (3,2) until the 27th iteration. At this stage, vertex A has a higher function value than vertices B and C, and A is consequently reflected through to D. D also has a larger function value than B and C and so, at the next iteration, is reflected back to A. The simplex therefore makes no further progress toward the minimum and oscillates repeated between ABC and BCD.

A variety of rules have been developed to control the movement and adaptation of the simplex, of which the most famous set is due to Nelder and Mead (Olsson and Nelson, 1975). The Nelder–Mead simplex procedure has been successfully used for a wide range of optimization problems and, due to its simple implementation, is amongst the most widely used of all optimization techniques. Importantly for the current application, simplex optimization is a black-box technique since it uses only the comparative values of the function at the vertices of the simplex to advance the position of the simplex, and it therefore requires no knowledge of the underlying mathematical function. It is also well suited to the optimization of expensive functions since as few as one new measurement is needed to advance the simplex one step. In its usual form, simplex optimization is suitable only for unconstrained optimization, but effective constrained versions have also been developed (Parsons et al., 2007;

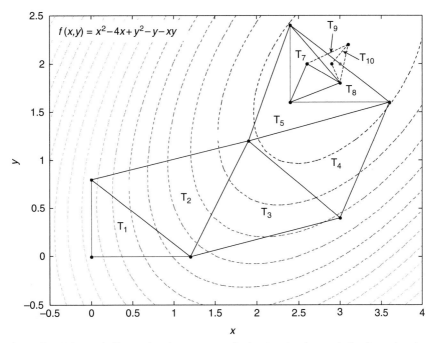

$$f(x,y) = x^2 - 4x + y^2 - y - xy$$

Figure 17 Schematic illustrating the concept of adaptive simplex optimization using the Nelder–Mead algorithm described in Olsson and Nelson (1975). The simplex initially expands in size and so makes rapid progress toward the minimum. It then contracts repeatedly, allowing it to converge on the minimum at (3,2).

Subrahmanyam, 1989). It can therefore be successfully used for constrained optimization of expensive black-box functions. Where it is less effective, however, is in the treatment of noisy functions. If the underlying noise-free function has similar values at the vertices, then the effect of the added noise may be sufficient to change the apparent ordering of the vertices, causing the optimization routine to make a bad decision such as unnecessary contraction (a process that should only happen when the simplex is close to the true minimum). The effect of such a contraction is to bring the vertices closer together, which means the values of the noise-free function at the vertices will be even closer and the distorting effect of noise even worse. The routine may therefore make a series of erroneous decisions that cause it to collapse to a point away from the true minimum.

The key weakness of the standard simplex method is that it only uses information about the function values at the vertices of the most recent simplex, and it entirely disregards data obtained at earlier stages in the optimization. The rejection of historical data as worthless is clearly naïve, but until recently, it has been difficult to see how such data could be straightforwardly incorporated into a simplex scheme (in which all decisions are based solely on the ranking of the vertices in the current simplex). This problem was solved in 2006 by Martinez (2005), who, building on earlier work by Anderson et al. (2000), proposed a statistical means of ranking the vertices. The detailed mathematical foundations need not concern us here, and it will suffice to say that the approach uses a statistical figure of merit—known as an optimality coefficient—whose value decreases monotonically from zero to minus one as the minimum is approached. In the scheme proposed by Martinez, the simplex optimization is implemented in the usual way, subject to the one change that vertices are ranked according to their relative optimality coefficients (instead of their relative function values). Importantly, the optimality coefficient is calculated using both the function values at the vertices and all historical data, and so the algorithm gains a more global view of the terrain. The statistical simplex approach is far less sensitive to the effects of noise and is therefore in turn less susceptible to becoming trapped at fictitious minima.

The application of the statistical simplex to nanoparticle synthesis is illustrated in Figure 18 using a simple one-dimensional example, in which the statistical simplex algorithm aimed to control the emission wavelength by varying the total flow rate ($F_{tot} = F_A + F_B$) of the injected precursors. (The ratio of the injection rates of the Cd and Se precursors was set equal, that is, $F_A = F_B$, to ensure a constant ratio of $Cd:Se$ in the reaction mixture.) The target wavelength was set to 540 nm, and the vertices of the initial simplex were arbitrarily set to 25 and 35 $\mu l\,min^{-1}$. The variation of the figure of merit, the flow rate, and the peak wavelength are shown as a function of measurement number in Figures 18a and b. In the initial stages of the optimization, the algorithm has no historical data on which to base its decisions and the flow rate oscillates wildly as the simplex reflects backward and forward. The algorithm gradually builds up a picture of the terrain, however, and as it does so the flow rate gradually converges to 5.75 $\mu l\,min^{-1}$, causing the figure of merit to reduce in magnitude toward zero and the peak wavelength to "home-in" on the target of 540 nm. The target is reached after approximately forty measurements. The data in Figure 18b reveal an approximate mirror symmetry between the flow rate F and the peak wavelength λ_{max}. This arises because slower flow rates imply longer reaction times which result in larger particles with longer peak emission wavelengths (and vice versa). In Figure 18c, we show the emission spectra corresponding to the first and last measurements, together with an interim spectrum obtained at measurement ten. The convergence on the target wavelength is clear. (The increase in intensity with measurement number is due to the increasing size of the particles as the emission moves to longer wavelengths, which reduces the quenching effects of surface defects.)

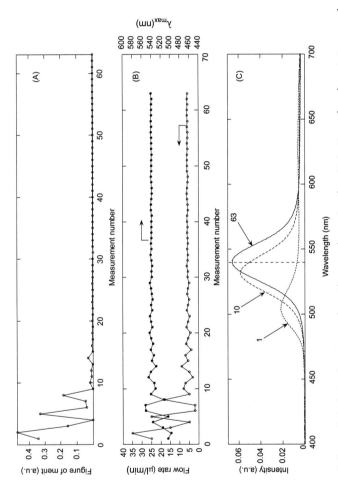

Figure 18 Application of the statistical simplex approach to the one-dimensional optimization of peak emission wavelength, using total flow rate as the sole reaction variable. (A) Variation of the figure of merit with measurement number. (B) Variation of the flow rate and peak wavelength with measurement number. (C) Emission spectra at various stages in the optimization, that is, the initial, tenth, and final measurements. The peak emission wavelength moves progressively closer to the target of 540 nm as the search proceeds.

The above example is intended mainly for the purposes of illustration, and it is arguable that one could just as easily identify the optimum flow rate by simply tweaking the flow rate manually. The automated approach is more useful for higher dimensional problems where the product can be controlled by multiple reaction conditions, for example, by varying the two injection rates of the precursors together with the temperature. In this case, multiple combinations of reaction conditions may yield the same emission wavelength since a faster flow rate (which reduces the reaction time) can, for example, be compensated by a higher reaction temperature (which increases the reaction rate). However, although multiple reaction conditions may yield particles with the same emission wavelength, the emission spectra are liable to have different line widths and intensities. In general one prefers nanoparticles that exhibit sharp intense emission, and it is interesting to ask whether one could optimize the peak wavelength subject to an additional requirement that the particles should, say, emit as brightly as possible. The optimization of multiple properties is known variously as multi-attribute optimization, multi-objective optimization, and decision theory, and is a subject of intense practical interest. There are various ways in which multi-attribute optimization can be carried out, but we will consider here the use of multi-attribute utility functions (MAUFs), that is, utility functions that combine multiple attribute values into a single figure of merit that can then be optimized in the usual way. MAUFs are most straightforwardly created by combining a number of single attribute utility functions (SAUFs) using, for example, a weighted sum. Again, we will not dwell on the mathematical details but simply note that the MAUFs are typically defined to have a value of unity when the multiple attributes all have their worst possible values which reduce to zero if and when they all have their best possible values.

The simultaneous optimization of peak wavelength and intensity is illustrated in Figure 19 for a target wavelength of 550 nm, using the two precursor injection rates as reaction variables. The variation of the figure of merit, the peak wavelength, and the peak intensity with measurement number is shown in Figures 19a–c, respectively. The reaction conditions sampled by the simplex algorithm are shown in Figure 19d, where the vertices of the starting simplex are marked out by a triangle and the final reaction condition is denoted by a grey circle at (3 ml/min, 1 ml/min). The interior region defined by the black dotted lines denotes the constrained parameter space. Figure 19a shows a steady decrease in the value of the figure of merit from an initial value of 0.37 to a final value of 0.2 due to an improvement in both the peak wavelength and the intensity. It is interesting to note that a close match to the wavelength is first reached at measurement 12, but continued searching leads to particles that emit with substantially increased intensity. The emission spectra obtained at measurements 12, 50, and 83 are compared in

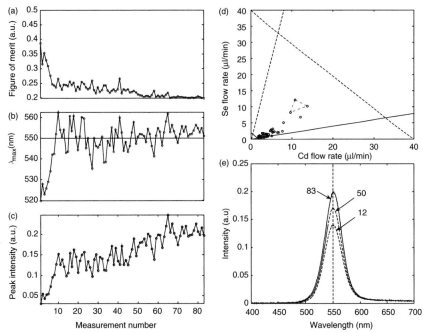

Figure 19 Application of the statistical simplex approach to the simultaneous optimization of peak emission wavelength and intensity, using the individual injection rates of the CdO and Se precursor solutions as the reaction variables. (A) Variation of the figure of merit with measurement number. (B) Variation of the peak wavelength with measurement number. (C) Variation of the intensity with measurement number. (D) Plot showing the conditions sampled by the statistical simplex routine. The initial simplex is denoted by the triangle and the final reaction conditions by the grey circle at (3 ml/min, 1 ml/min). The trapezoidal region defined by the four overlapping dotted lines defines the constrained reaction space. The simplex was only permitted to sample reaction conditions inside this space. (E) Emission spectra at various stages in the optimization, that is, at the 12th, 50th, and 83rd measurement. The peak wavelength shows a close match to the target wavelength in each case but the intensity increases progressively as the search continues, indicating an improvement in particle quality.

Figure 19d, and it is evident that, despite their similar peak wavelengths, the final particles are evidently the most intense.

The above results confirm the feasibility of automating the process of nanoparticle synthesis using simplex-based optimization routines. The approach we outline provides a promising starting point for developing more sophisticated systems that can be used to control a whole variety of nanoparticle properties, including shape, chemical composition, crystal structure, optical properties, and chemical reactivity. In passing, we note that—since the simplex converges on the optimum though a series of reflections in the general direction of improvement—there is no guarantee it will find the global optimum, which may be located in a distant unexplored region of the parameter space. To stand a better chance of finding the global optimum, different optimization routines are required that work by sampling at discrete locations throughout the parameter space. The search process is divided into two phases: (1) local searching (i.e., solution refinement) in the vicinity of identified optima and (2) global searching in hitherto unexplored regions of the parameter space where superior optima might potentially exist. The interested reader is referred to Krishnadasan et al. (2007) for a discussion of how global optimizers can be applied to nanoparticle synthesis.

5. PROCESS CONTROL

There are two key steps to any practical optimization procedure: firstly, as described in the previous section, the optimal reaction conditions should be identified as efficiently as possible to enable swift production of the desired product, and secondly, having found the (initially) optimal conditions, the reaction conditions should then be regularly updated to compensate for any changes (drift) in the system due, for example, to degradation of the precursor materials or furring of the channel walls. The effects of drift are essentially due to progressive changes in hidden variables over which the user has no control. To compensate for these changes, the active variables must be adjusted accordingly. Hence if, for example, aggressive furring of the channels were to occur, this might significantly reduce the cross-sectional area of the channels and so reduce the residence time for a given flow rate. This could be mitigated either by reducing the total flow rate (to increase the residence time) or increasing the temperature (to increase the reaction rate) or by making appropriate changes to both.

In most cases, the detailed effects of system drift are unpredictable, and it is again important to use black-box algorithms that are able to make adjustments to the reaction conditions on the basis of measured data only without the need for any underlying process model. In order for such a black-box algorithm to determine how to update the reaction conditions, it must constantly dither the reaction conditions about (what it perceives to be) the current optimum. In other words, the algorithm must constantly switch between similar, but slightly different, reaction conditions that yield end products of marginally varying quality. The algorithm can then compare the quality of the particles so obtained and use this information to update the reaction conditions in an appropriate manner. The constant perturbing of the reaction conditions is a necessary activity for tracking the system drift but unavoidably compromises the instantaneous quality of the product since the system frequently finds itself operating at (slightly) non-optimal conditions.

To be effective, the perturbations should be sufficiently small that the product is always of an acceptable quality yet sufficiently large that meaningful comparisons can be made between the different conditions.

Xiong and Jutan have recently reported a real-time algorithm for tracking system drift using a simple adaptation of the reflective simplex approach outlined above (Xiong and Jutan, 2003). In essence, the simplex undergoes repeated reflections that allow it to track the moving optimum. If the optimum is static, the simplex reflects backward and forward and hence remains more or less anchored to the optimum. If the optimum starts to move, however, the repeated reflections drive the simplex in the direction of the drifting optimum.

As noted above, the effects of system drift may be attributed to hidden variables over which the algorithm has no control. In reality, for CdSe nanoparticles synthesized in the manner described above, system drift is a very slow process that occurs over a timescale of days. In order to accelerate the effects of drift and to pose a challenging test for our algorithm, we engineered an artificial situation in which the reaction temperature was programmed to drift linearly with time from 180 to 230 °C over a period of 6 h. (In ordinary circumstances, this would be expected to induce a significant red-shift in the peak emission wavelength.) The system was set to an initial total flow rate of $12 \, \mu l \, min^{-1}$, corresponding to particles that initially emitted at 510 nm. The evolution of spectra under fixed and dynamically updated flow-rate conditions was then compared. The upper line in Figure 20 indicates the progressive change in peak wavelength observed over the 6-h period when the flow rate was fixed at $12 \, \mu l \, min^{-1}$. The peak wavelength increases almost linearly with time, indicating the formation of larger particles with red-shifted emission (due to the increased temperature-dependent reaction rate, c.f. Figure 10). The peak wavelength shifts by almost 30 nm by the end of the 6-h period. The lower line indicates the variation in the peak wavelength when the flow rate was dynamically updated by the drift-compensating algorithm. The peak wavelength shows an increased volatility due to the constant dithering of the reaction conditions and exhibits a few undesirable outliers in the early stages of the optimization (see circled data points). However, the baseline is effectively static, which indicates that the algorithm is successfully tracking the drifting optimum, and after 6 h the peak wavelength is still approximately 510 nm. The evolution of the flow rate is shown in Figure 21. The flow rate tracks (albeit in a somewhat hap-hazard fashion) the temperature, ensuring that the reaction time decreases sufficiently to offset the increase in temperature. The evolution of the emission spectra with time are shown in Figure 19a and b. The non-stabilized system gives rise to a progressive red shift in the particle emission with time, accompanied by a progressive reduction in the intensity of the defect emission (due to the larger size of the particles). The dynamically stabilized system

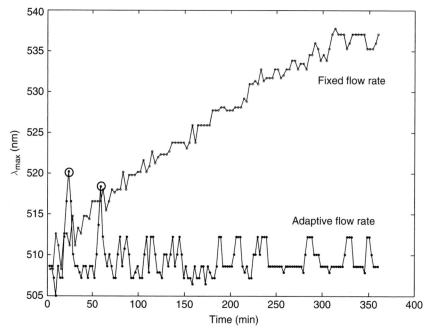

Figure 20 Application of the dynamic simplex to the compensation of system-drift. An artificial example is considered here in which the temperature is ramped linearly with time and the simplex aims to compensate for the changes in the reaction temperature by modifying the flow rate accordingly. The plot compares the change in the peak wavelength when the flow rate is held fixed at its initial value of 12 µl min^{-1} and when it is adapted dynamically by the simplex algorithm. In the former case, the peak wavelength increases steadily with time due to the increasing temperature which increases the growth rate of the particles. In the latter case, the peak wavelength remains fairly close to its initial value of 508 nm.

however shows a much smaller variation in the emission spectra. The remarkable effectiveness of the dynamic stabilization is particularly evident from a comparison of Figure 22c and d, which compare the emission spectra at time 0 and after 6 h for the two cases. The dynamically stabilized system shows a remarkably small change in the emission spectrum, indicating the efficacy of the simplex technique.

6. THE APPLICATION OF AUTOMATED MICROREACTORS IN NANOTOXICOLOGY

In the above sections, we have outlined our recent work in the area of nanoparticle automation and have demonstrated how simple algorithms can be developed to address the two key challenges in nanoparticle

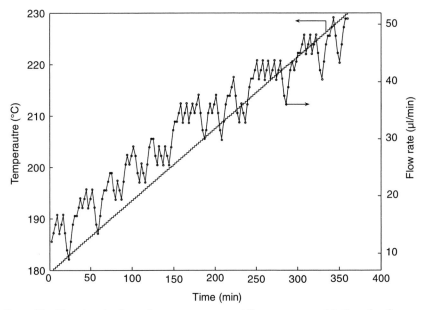

Figure 21   Plot showing how the temperature and flow rate vary with time for the situation considered in Figure 20. The flow rate tracks (albeit in a somewhat hap-hazard fashion) the temperature, ensuring that the reaction time decreases as the temperature increases and that the particles therefore continue to emit with the same approximate peak wavelength.

automation—*fast identification of the optimal reaction conditions and compensation for system-drift*. In this final section, we will spend a few paragraphs outlining how such automated systems might usefully be applied in the specific area of nanotoxicology. In the case of conventional macroscale materials, toxicity is determined primarily by chemical composition. The particle dimensions do not strongly influence the physicochemical properties of the materials themselves, and hence they tend to influence toxicity only indirectly by in essence determining how easily a given particle can reach a susceptible location in the body. In the case of asbestos, for example, the diameter of the fibers determines their ability to penetrate the lung and reach the alveoli where they provoke an inflammatory response that causes deposition of fibrous tissue (and in so doing reduces lung capacity). The thin linear "amphibole" fibers pictured in Figure 23a are able to penetrate the lungs more readily than the curved "serpentine" fibers pictured in Figure 23b and so typically cause more damage. In the case of nanoscale materials, however, the situation is more complex because the particle dimensions also have a strong influence on the physicochemical properties of the particles. In principle, even small variations

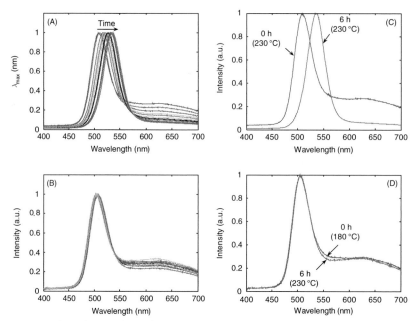

Figure 22 Plots showing how the emission spectra vary with time for the situation considered in Figure 20. (A) The evolution of the spectra for a fixed flow rate of 12 µl min⁻¹. (B) The evolution of the spectra when the flow rates are dynamically updated by the simplex algorithm. (C) Comparison of the initial and final spectra for a fixed flow rate of 12 µl min⁻¹. (D) Comparison of the initial and final spectra when the flow rates are dynamically updated by the simplex algorithm.

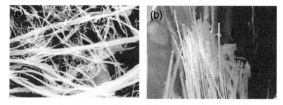

Figure 23 Two varieties of asbestos fibers: (A) amphibole and (B) serpentine. The thin amphibole fibers present a larger toxicological threat due to their greater ability to penetrate lung tissue and reach the alveoli.

in the size and shape of the particles may lead to large variations in their properties and, by implication, their toxicity. Hence, in order to develop a detailed understanding of the toxic effects of a particular nanomaterial, it is important to study its toxicological behavior over the full range of sizes and shapes that are liable to be encountered in practical situations (rather than a few discrete cases as is usually sufficient for toxicological studies of macroscale particles). Furthermore, a number of other factors are likely to influence toxicity, including crystallinity, surface functionalization, surface charge (zeta potential), hydrophilicity, and propensity for aggregation or agglomeration. The diverse range of factors that can influence the toxicity of nanomaterials presents a formidable challenge for toxicologists, and there is a growing need for new tools and techniques that can be used to analyze the toxicity of nanomaterials.

Conventional macroscale toxicology makes heavy use of libraries of reference materials (RMs)—that is, standard materials with stable well-defined properties that are prepared and handled according to precise protocols. RMs may be investigated by multiple researchers in the confidence that they are all looking at essentially the same materials, enabling meaningful comparisons to be made between their research findings. There is now an urgent need to develop similar libraries for nanotoxicology, in which commonly used nanomaterials are provided in as wide a variety of sizes, shapes, and surface functionalizations as possible. A number of recent studies have investigated the feasibility of developing such libraries, and there is an emerging consensus about the requirements of such libraries. Firstly, RMs should be available for all the most commonly used nanomaterials, namely, metals, metal alloys, metal oxides, quantum dots, and fullerenes. Secondly, the RMs should be provided in a range of monodisperse sizes and shapes. Thirdly, the RMs should be of verifiable quality with minimal batch-to-batch variation between nominally identical materials. And finally, the RMs should be available in kilogram quantities to ensure sufficient material is available for worldwide studies. In addition, the studies drew attention to one issue of particular concern, namely, the significant differences in reactivity (and hence potential toxicity) that frequently exist between freshly prepared and stored particles. This issue throws into serious doubt the viability of the conventional RM approach—in which standardized materials are prepared in centralized facilities and then transported to their intended location of use—since the particles are liable to "age" substantially in transit. Ideally, the nanomaterials should be prepared on demand at the point of use, but this is inconvenient, potentially dangerous, and requires a level of synthetic expertise that the typical practicing toxicologist does not have. The method outlined in this chapter, however, offers a possible solution. The automated systems described above could be readily integrated into a simple self-contained desktop machine that produces nanomaterials on demand to a well-defined and verifiable specification. In this way, nanomaterials could be prepared as and when required, enabling reliable toxicological studies to be undertaken on both fresh and aged samples. Moreover, the high levels of process control afforded by the microfluidic format enable systematic studies in which key nanomaterials properties—size, shape, chemical composition, zeta potential, etc.—may be varied with ease. At the time of writing, a wide variety of nanomaterials have been successfully prepared in microreactors, including CdSe, CdS TiO2, Ag, Au, and Co to name but a few. In fact, of the key families of nanomaterials identified above, only fullerenes have yet to be successfully synthesized in a microfluidic format (due to the normal use of combustion-based preparation routes). The automated microfluidic route is therefore an attractive means of making most of the commonly used families of nanomaterials. Importantly, it satisfies most of the other requirements of RMs noted above: the high levels of process control enable the preparation of size-monodisperse series of nanomaterials and the use of online monitoring allows quality assurance and minimizes batch-to-batch variation.

The major objection to using the microfluidic approach might seem to be inadequate materials throughput. In fact, despite the small size of the reaction channels, surprisingly large quantities of materials can be prepared using microfluidic devices. In the case of CdSe quantum dots, for example, the nanoparticles can be prepared using flow rates of several hundred microliters per minute which, for typical precursor concentrations of $10\,\mu g\,\mu l^{-1}$ implies production rates of order $1\,mg\,min^{-1}$. Moreover, due to the small size of the channels, it is possible to integrate multiple channels onto a single microchip, enabling the syntheses to be carried out in parallel. This approach is called "scale-out" in contrast to the usual process of "scale-up" and has the important advantage that all channels can be fabricated with identical dimensions, and the increased throughput can therefore be achieved without the need to reformulate the chemical reactions in any way (as is invariably necessary for scale-up). In a small multichannel system containing just twenty parallel reactors, production rates of about $1\,gh^{-1}$ are easily achievable which is sufficient for most practical applications. Hence, despite the small size of the microreactors, an appreciable amount of material can be produced in a relatively short period of time.

In our current work, we are aiming to develop parallel systems of this nature for a variety of applications where on-demand production of high-quality nanoparticles is required.

7. CONCLUSIONS

It is fair to say that over the last 8 years there has been a significant growth in microfluidic-based methods for synthesizing nanoscale materials. Significantly, such materials are of a quality that matches or exceeds materials produced using traditional bulk methods. Furthermore, the continued development of integrated multicomponent systems will create invaluable tools for improving both the properties and yields of this important class of materials.

More specifically, the described approaches for reaction automation represent an important first step toward both simplifying and automating the process of nanoparticle production, but there is still considerable scope for improvement. To date, we have applied our approaches to a relatively small number of nanoparticle syntheses using very simple chemical routes. However, some extremely sophisticated chemical syntheses have been developed in recent years that are capable of producing near defect-free nanoparticles of extremely high monodispersity, and it will be important to adapt and apply our automation procedures to these synthesis routes also. In addition, although in this chapter we have focused on the synthesis of fluorescent quantum dots, the general strategy we describe has wider applicability to any particles whose properties can be monitored (directly or indirectly) inline. In the case of nonfluorescent dots and rods, dynamic light scattering is likely to provide an especially effective tool for controlling the size, shapes, and dispersity of the particles. As mentioned above, careful design of the control algorithms is crucial to achieving effective automation, and considerable scope exists for achieving further efficiency and control through improved algorithm design. Notwithstanding these challenges, we consider the general approach outlined above to offer a powerful route to the automated production of optimized nanoparticles which has the potential to transform the efficacy of nanoparticle synthesis in terms of control, yield, and ease-of-use.

ACKNOWLEDGEMENT

We acknowledge financial support from the European Commission (NanoReTox, http://www.nanoretox.eu) and the Royal Society under its Industry Fellowship scheme.

REFERENCES

Alivisatos, A. P. *Science* **271**, 933–937 (1996).
Anderson, B., Moore, A., and Cohn, D. A. "Presented at International Conference on Machine Learning 2000", Stanford University, CA, USA (2000).
Boleininger, J., Kurz, A., Reuss, V., and Sonnichsen, C. *Phys. Chem. Chem. Phys.* **8**, 3824–3827 (2006).
Chan, E. M., Alivisatos, A. P., and Mathies, R. A. *J. Am. Chem. Soc.* **127**, 13854–13861 (2005).
Chan, E. M., Mathies, R. A., and Alivisatos, A. P. *Nano Lett.* **3**, 199–201 (2003).
Cottam, B. F., Krishnadasan, S., deMello, A. J., deMello, J. C., and Shaffer, M. S.P. *Lab. Chip* **7**, 167–169 (2007).
deMello, A. J. *Nature* **442**, 394–402 (2006).
DeMello, J., and DeMello, A. *Lab Chip* **4**, 11N–15N (2004).
Donega, C. D., Hickey, S. G., Wuister, S. F., Vanmaekelbergh, D., and Meijerink, A. *J. Phys. Chem. B* **107**, 489–496 (2003).
Dumais, P., Callender, C. L., Ledderhof, C. J., and Noad, J. P. *Appl. Opt.* **45**, 9182–9190 (2006).

Dushkin, C. D., Saita, S., Yoshie, K., and Yamaguchi, Y. *Adv. Colloid Interface Sci.* **88**, 37–78 (2000).

Edel, J. B., Fortt, R., deMello, J. C., and deMello, A. J. *Chem. Commun.* 1136–1137 (2002).

Euliss, L. E., DuPont, J. A., Gratton, S., and DeSimone, J. *Chem. Soc. Rev.* **35**, 1095–1104 (2006).

Green, M. *Angew. Chem. Int. Ed.* **43**, 4129–4131 (2004).

Han, G., Ghosh, P., and Rotello, V. M. *Nanomedicine* **2**, 113–123 (2007).

Hung, L.-H., Choi, K. M., Tseng, W.-Y., Tan, Y.-C., Shea, K. J., and Lee, A. P. *Lab Chip* **6**, 174–178 (2006).

Iliescu, C. *Informacije Midem-J. Microelectron. Electron. Compon. Mater.* **36**, 204–211 (2006).

Kawazoe, T., Yatsui, T., and Ohtsu, M. *J. Non-Crystall. Solids* **352**, 2492–2495 (2006).

Khan, S. A., Gunther, A., Schmidt, M. A., and Jensen, K. F. *Langmuir* **20**, 8604–8611 (2004).

Klostranec, J. M., and Chan, W. C. W. *Adv. Mater.* **18**, 1953–1964 (2006).

Kohler, J. M., Wagner, J., and Albert, J. *J. Mater. Chem.* **15**, 1924–1930 (2005).

Kolda, T. G., Lewis, R. M., and Torczon, V. *Siam Rev.* **45**, 385–482 (2003).

Koo, J. M., and Kleinstreuer, C. *J. Micromech. Microeng.* **13**, 568–579 (2003).

Krishnadasan, S., Brown, R., deMello, A. J., and deMello, J. C. *Lab Chip* **7**, 1434–1441 (2007).

Krishnadasan, S., Tovilla, J., Vilar, R., deMello, A. J., and deMello, J. C. *J. Mater. Chem.* **14**, 2655–2660 (2004).

La Mer, V. K., and Dinegar, R. H. *J. Am. Chem. Soc.* **72**, 4847–4854 (1950).

Malik, M. A., O'Brien, P., and Revaprasadu, N. *Phosphorus Sulfur and Silicon and the Relat. Elem.* **180**, 689–712 (2005).

Manz, A., Harrison, D. J., Verpoorte, E. M. J., Fettinger, J. C., Paulus, A., Ludi, H., and Widmer, H. M. *J. Chromatogr.* **593**, 253–258 (1992).

Martinez, E. C. *Indus. Eng. Chem. Res.* **44**, 8796–8805 (2005).

Masala, O., and Seshadri, R. *Annu. Rev. Mater. Res.* **34**, 41–81 (2004).

Matsui, I. *J. Chem. Eng. Jpn* **38**, 535–546 (2005).

Medina, C., Santos-Martinez, M. J., Radomski, A., Corrigan, O. I., and Radomski, M. W. *Br. J. Pharmacol.* **150**, 552–558 (2007).

Milliron, D. J., Hughes, S., and Alivisatos, A. P. *Abstr. Pap. Am. Chem. Soc.* **227**, U274–U274 (2004).

Millman, J. R., Bhatt, K. H., Prevo, B. G., and Velev, O. D. *Nat. Mater.* **4**, 98–102 (2005).

Murray, C. B., Norris, D. J., and Bawendi, M. G. *J. Am. Chem. Soc.* **115**, 8706–8715 (1993).

Nakamura, H., Yamaguchi, Y., Miyazaki, M., Maeda, H., Uehara, M., and Mulvaney, P. *Chem. Commun.* 2844–2845 (2002).

Nightingale, A. M., and de Mello, J. C. *Chem. Phys. Chem.* **10**, 2612–2614 (2009).

Olsson, D. M., and Nelson, L. S. *Technometrics* **17**, 45–51 (1975).

Ozin, G., and Arsenault, A. *Nanochem. A Chem. Approach Nanomater.* (2005).

Park, J., An, K. J., Hwang, Y. S., Park, J. G., Noh, H. J., Kim, J. Y., Park, J. H., Hwang, N. M., and Hyeon, T. *Nat. Mater.* **3**, 891–895 (2004).

Parsons, D. J., Green, D. M., Schofield, C. P., and Whittemore, C. T. *Biosyst. Eng.* **96**, 257–266 (2007).

Peng, Z. A., and Peng, X. G. *J. Am. Chem. Soc.* 123, 183–184 (2001).

Pradhan, N., Katz, B., and Efrima, S. *J. Phys. Chem. B* **107**, 13843–13854 (2003).

Rao, C. N. R., Agrawal, V. V., Biswas, K., Gautam, U. K., Ghosh, M., Govindaraj, A., Kulkarni, G. U., Kalyanikutty, K. R., Sardar, K., and Vivekchandi, S. R. C. *Pure Appl. Chem.* **78**, 1619–1650 (2006).

Rhyner, M. N., Smith, A. M., Gao, X. H., Mao, H., Yang, L. L., and Nie, S. M. *Nanomedicine* **1**, 209–217 (2006).

Shalom, D., Wootton, R. C. R., Winkle, R. F., Cottam, B. F., Vilar, R., deMello, A. J., and Wilde, C. P. *Mater. Lett.* **61**, 1146–1150 (2007).

Shestopalov, I., Tice, J. D., and Ismagilov, R. F. *Lab Chip* **4**, 316–321 (2004).

Song, Y. J., Modrow, H., Henry, L. L., Saw, C. K., Doomes, E. E., Palshin, V., Hormes, J., and Kumar, C. *Chem. Mater.* **18**, 2817–2827 (2006).

Subrahmanyam, M. B. *J. Optim. Theory Appl.* **62**, 311–319 (1989).

Takagi, M., Maki, T., Miyahara, M., and Mae, K. *Chem. Eng. J.* **101**, 269–276 (2004).

Uehara, M., Nakamura, H., and Maeda, H. *J. Nanosci. Nanotechnol.* **9**, 577–583 (2009).

Vilkner, T., Janasek, D., and Manz, A. *Anal. Chem.* **76**, 3373–3385 (2004).

Wagner, J., and Kohler, J. M. *Nano Lett.* **5**, 685–691 (2005).

Wang, H. Z., Li, X. Y., Uehara, M., Yamaguchi, Y., Nakamura, H., Miyazaki, M. P., Shimizu, H., and Maeda, H. *Chem. Commun.* 48–49 (2004).

Wang, H. Z., Nakamura, H., Uehara, M., Yamaguchi, Y., Miyazaki, M., and Maeda, H. *Adv. Funct. Mater.* **15**, 603–608 (2005).

Xia, D. Y., Li, D., Ku, Z. Y., Luo, Y., and Brueck, S. R.J. *Langmuir* **23**, 5377–5385 (2007).

Xiong, Q., and Jutan, A. *Chem. Eng. Sci.* **58**, 3817–3828 (2003).

Xue, Z. L., Terepka, A. D., and Hong, Y. *Nano Lett.* **4**, 2227–2232 (2004).

Yang, H., Luan, W., Wan, Z., Tu, S-t., Yuan, W.-K., and Wang, Z. M. *Cryst. Growth Des.* **9**, 4807–4813 (2009).

Yen, B. K. H., Günther, A., Schmidt, M. A., Jensen, K. F., and Bawendi, M. G. A. *Angew. Chem. Int. Ed.* **44**, 5447–5451 (2005).

Yen, B. K. H., Stott, N. E., Jensen, K. F., and Bawendi, M. G. *Adv. Mater.* **15**, 1858–1862 (2003).

Yordanov, G. G., Gicheva, G. D., Bochev, B. H., Dushkin, C. D., and Adachi, E. *Coll. Surf. A-Physicochem. Eng. Asp.* **273**, 10–15 (2006).

Note: The letters 'f' and 't' following locators refer to figures and tables respectively.

CONTENTS OF VOLUMES IN THIS SERIAL

Volume 11 (1981)

Jean-Claude Charpentier, *Mass-Transfer Rates in Gas–Liquid Absorbers and Reactors*
Dee H. Barker and C. R. Mitra, *The Indian Chemical Industry—Its Development and Needs*
Lawrence L. Tavlarides and Michael Stamatoudis, *The Analysis of Interphase Reactions and Mass Transfer in Liquid–Liquid Dispersions*
Terukatsu Miyauchi, Shintaro Furusaki, Shigeharu Morooka, and Yoneichi Ikeda, *Transport Phenomena and Reaction in Fluidized Catalyst Beds*

Volume 12 (1983)

C. D. Prater, J, Wei, V. W. Weekman, Jr., and B. Gross, *A Reaction Engineering Case History: Coke Burning in Thermofor Catalytic Cracking Regenerators*
Costel D. Denson, *Stripping Operations in Polymer Processing*
Robert C. Reid, *Rapid Phase Transitions from Liquid to Vapor*
John H. Seinfeld, *Atmospheric Diffusion Theory*

Volume 13 (1987)

Edward G. Jefferson, *Future Opportunities in Chemical Engineering*
Eli Ruckenstein, *Analysis of Transport Phenomena Using Scaling and Physical Models*
Rohit Khanna and John H. Seinfeld, *Mathematical Modeling of Packed Bed Reactors: Numerical Solutions and Control Model Development*
Michael P. Ramage, Kenneth R. Graziano, Paul H. Schipper, Frederick J. Krambeck, and Byung C. Choi, *KINPTR (Mobil's Kinetic Reforming Model): A Review of Mobil's Industrial Process Modeling Philosophy*

Volume 14 (1988)

Richard D. Colberg and Manfred Morari, *Analysis and Synthesis of Resilient Heat Exchange Networks*
Richard J. Quann, Robert A. Ware, Chi-Wen Hung, and James Wei, *Catalytic Hydrometallation of Petroleum*
Kent David, *The Safety Matrix: People Applying Technology to Yield Safe Chemical Plants and Products*

Volume 15 (1990)

Pierre M. Adler, Ali Nadim, and Howard Brenner, *Rheological Models of Suspenions*
Stanley M. Englund, *Opportunities in the Design of Inherently Safer Chemical Plants*
H. J. Ploehn and W. B. Russel, *Interations between Colloidal Particles and Soluble Polymers*

Volume 16 (1991)

Perspectives in Chemical Engineering: Research and Education
Clark K. Colton, *Editor*
Historical Perspective and Overview
L. E. Scriven, *On the Emergence and Evolution of Chemical Engineering*
Ralph Landau, *Academic—industrial Interaction in the Early Development of Chemical Engineering*